ScanCAD Edition

Ng Keng Tiong
with
Bill Loving & Jeff Rupert

Copyright © 2017 Ng Keng Tiong. All rights reserved.

Cover design by the author.

Products and services mentioned in this book are trademarks or registered trademarks of their respective companies. All trademarks and registered trademarks are the property of their respective holders.

No part of this book may be reproduced in any form, or stored in a database or retrieval system, or transmitted or distributed in any form, by any means, electronic, mechanical, photocopying, recording, or otherwise, without the prior written permission of the author.

LIMIT OF LIABILITY AND DISCLAIMER OF WARRANTY

The information, examples, illustrations, documentation, and other references in this book are provided "as is", without warranty of any kind, expressed or implied, including without limitation any warranty concerning the accuracy, adequacy, or completeness of the material or the results obtained from using the material. Neither the publisher nor the author shall be responsible for any claims attributable to errors, omissions, or other inaccuracies in the material in this book. In no event shall the publisher or author be liable for direct, indirect, special, incidental, or consequential damages in connection with, or arising out of, the construction, performance, or other use of the materials contained herein.

ISBN   : 1977549039
ISBN-13 : 978-1977549037

To all the aspiring engineers out there who, at one time or another in their career endeavors, have embarked or are contemplating taking up this challenging but rewarding journey of doing PCB reverse engineering.

May the resources in this book and the shared experiences of fellow engineers offer you greater insights and directions, and make your brave ventures into the world of PCB-RE a delightful and exciting one on this road that is less travelled.

Acknowledgements

Special thanks to the staff of ScanCAD International Inc. who in one way or another, have contributed their time and effort to go through the initial draft for the ScanCAD chapter. I would also like to thank Lee Foreman and Sharon Lujan for proofreading the final draft of this book and offering helpful suggestions in their findings.

FOREWORD

It is inconceivable that a book of this nature could originate from two locations, separated by a vast geographical distance——one near equatorial Singapore and another high up in the Rocky Mountains of Colorado——and by two parties with completely different perspectives, coming together to address the unusual and rare topic of PCB Reverse Engineering (PCB-RE).

Keng Tiong (KT) took up PCB-RE out of necessity in his work, adopting a hands-on manual approach to keep his customers' equipment alive and functional; this is documented in his enjoyable and easy to read book, *The Art of PCB Reverse Engineering*. ScanCAD International team's approach to PCB-RE, however, has a much longer history and evolutionary process. In the late 1980s, it filled the needs of the PCB industry to convert vast PCB design archives from plain, static image media (film, mylar, diazo, hand-taped artwork, micro-fiche, paper, etc.) into electronic data in standard CAD formats. Hence the name, ScanCAD——scan images to create CAD data. The initial processes were designed to digitize or vectorize PCB images on flat documents and import them into CAD/CAM environments for use in automated PCB fabrication and assembly operations, while safeguarding precious data at the same time.

Early generations of PCBs, forming a large pool of valuable intellectual property assets, were created pre-CAD and stored on deteriorating physical materials within archives under poor environmental control. These prized Intellectual Property (IP) were in danger of being lost but thankfully ScanCAD came into the picture. Vast amounts of PCB image data have since been successfully converted to vector data in over 48 countries; valuable data that would have otherwise been lost forever.

Over time, the PCB-RE needs have shifted from converting media archives to that of converting physical PCBs into CAD data. In many instances, the only remaining source of the IP is the actual PCB. ScanCAD systems were enhanced over the years to image the top and bottom of PCBs, incorporating new technologies to safely and reliably de-laminate PCBs to expose and retrieve the underlying layer designs, thereby saving the IP of the PCBs. Other technologies such as X-ray, CT-scan and Flying Probe Test (FPT) equipment have also been integrated into the PCB-RE work process with the goal of reliably reproducing PCB CAD data when the original source data has been lost or destroyed.

As KT stated in his first book, PCB-RE is about *unraveling the beauty of the original design* hence the 'art' of PCB-RE. The original design represents a valuable IP asset to the owner. When this design or asset is lost due to a computer virus, rogue employee, accident, vendor bankruptcy or a thousand other causes, the PCB can't be repaired or replaced. Virtually all of today's manufactured products depend heavily on electronics to run, without which our world would come to a complete standstill. Safeguarding the original design, therefore, is a top priority and the primary focus of this book.

Keeping existing legacy electronics operating smoothly not only reduces financial loss, management and operational stress in organizations, it also reduces environmental impact on our planet by reducing scrap and saving resources. The cost of discarding and scrapping major systems and products due to inability to repair or fabricate electronic replacement parts is both high and wasteful. Moreover, keeping electronics in legacy systems running can be measured in yet a different way—imagine the benefit of keeping legacy medical imaging systems operational for developing countries—it can be challenging or impossible to obtain replacement parts for such obsolete equipment that could save lives for years to come, for lack of a simple electronic part.

The same can be said about communications, transportation and other equipment, etc. It is one thing to discard or recycle a mobile phone or laptop PC that has become obsolete—products that are 99% electronic and low cost. It is entirely different to discard a medical X-ray, ultrasound or MRI machine, an aircraft, a naval vessel, a subway/train or air traffic control system, that is, equipment that have major mechanical components in addition to electronics. This is happening every day and with increasing frequency. Huge resources are wasted and opportunities lost by not keeping legacy systems running. In yet other cases, the safety of entire communities is put at risk because vital equipment is taken offline. Electricity, water and gas utilities are all controlled by electronics today. Most are legacy systems. Many are short of spare parts. The list goes on and on.

This book is intended to help bring hope and light to these PCB-RE situations:

- To help today's engineers keep electronic equipment running when data has been lost.
- To help entrepreneurs bring clever solutions to people in need.
- To help manage the precious resources of our planet.

Bottom line—be a beneficial presence in our world.

This book also honors ScanCAD's twenty-eight-year vision:

> *ScanCAD International improves the quality of life for customers, employees and the world by providing high quality, innovative solutions that simplify complex technology. ScanCAD accomplishes this in a positive, environmentally-conscious manner while having fun along the way!*

William F. Loving
President and CEO
ScanCAD International, Inc.
October 1, 2017

PREFACE

The only source of knowledge is experience.

Albert Einstein

I agree with Einstein. Acquiring knowledge through personal experience, however, can be costly and time-consuming. Learning from the valuable experiences of others can help offset the cost and reduce the time to acquire that knowledge. While engineering books of various disciplinary fields abound without the slightest sign of abating, in-depth coverage of niche topics like PCB reverse engineering (PCB-RE) are surprisingly rare, even though it is widely practiced within the repair and refurbish industries of old and obsolete PCBs and systems. A reader who bought my book, *The Art of PCB Reverse Engineering*, left this comment on the Amazon website:

> Reverse engineering PCBs is to electronic circuit boards what hacking is to computer software...something people do, but no one admits to doing it.
>
> Mr. Ng not only does it, but he lays out a very clear and systematic approach to the entire process. I am self-taught in this field, and it was a real joy to find that someone had actually written an entire book on the subject. I highly recommend this book to anyone interested in the topic.

Honestly, I did not expect such hearty endorsement from an international reader when I started out to write my very first engineering book, having sold more copies than I had anticipated in just two years and garnered mostly positive reviews online and through emails, expressing sincere appreciations and thanks, even though I am an obscured self-publish author without notable credentials or fame.

Such reactions from readers, engineers who are mostly self-taught in PCB-RE, seem to suggest that there is a need for more of such books to be written, for there can never be enough for lack of real substance and shared experiences as far as this coveted knowledge and closely guarded skillset among professional practitioners is concerned. While my first book addresses the challenges of doing PCB-RE using the manual approach, targeting mainly hobbyists and repair personnel who do not have the luxury of expensive equipment but are required to perform such tasks on an ad hoc basis at work, or simply to find out how a PCB works or why it failed, it only briefly touched on alternative options available to those who may want to explore further.

I reckoned there was a possibility of writing a sequel book to complement and complete the proverbial dots. After deliberating for two years, and more recently when I came across a review dated just October last year on the 0x90.se website, where the writer listed a few would-be-nice-to-have topics, I finally made up my mind to revisit the PCB-RE topic again, this time to give it a more thorough treatment.

As I went about to research and read the online published works of various authors who are experts in their own fields of PCB-RE, it became agonizingly clear that writing a book of this magnitude all by myself is impractical if not nearly impossible. That's when I remembered the adage: 'There is strength in unity.' and decided that the best cause of action is to invite those with the relevant experience and expertise to contribute their valuable knowledge on this topic of PCB-RE.

One of the first few people I approached who responded early and positively is Jeff Rupert, an engineer from ScanCAD International, Inc. He not only saw the potential of such a cooperative venture but persuaded Bill Loving, a fellow engineer and the CEO of the company, to become interested and involved as well. Their support and encouragement has enabled me to go further than what I first intended, and while the book is still work in progress as of this writing, I decided to return them a favor by coming out with a special edition of the book, dedicated to ScanCAD as a gesture of my heart-felt appreciation.

It is my hope that the insights offered within its pages will provide readers with a better idea about the PCB-RE subject, as well as the practical know-how from the company that produces the world's #1 selling PCB re-engineering systems since 1990.

Ng Keng Tiong
September 23, 2017

TABLE OF CONTENTS

1. FUNDAMENTALS

The Basics of PCB-RE   15

    What is PCB-RE?
    Why PCB-RE?
    Anatomy of a PCB
    Understanding the PCB-RE Process
    Evaluating Your Chances
    Going a Little Further...
    A Rare Interview
    Why PCB-RE? (Again)
    Anti-Tampering Measures
    But is it Good Enough?
    A Little Interesting Anecdote
    The Defining Moment

2. TOOLS & TECHNIQUES

ScanCAD: The Art of Perfect PCB-RE   49

    A Legacy Crisis
    PCB-RE Considerations
    Legacy PCB Data
    Manual Probing
    Optical Imaging
    Bare Board FPT
    Bare Board BON*
    Loaded Board FPT
    Optical Scanning and X-ray Imaging
    Verification & Independent Validation Technique
    Netlist Data Conversion
    Schematic Data Conversion
    Conclusion

* BON – Bed-of-nails

3. VENDOR INFORMATION

Equipment & Software   87

    ScanCAD International, Inc.
    ScanFAB
    Precision Material Removal System
    FPT Software Module
    ScanPLACE
    ConvertPLUS ARE
    Schematic Generation Module
    ScanINSPECT
    ProWorks Electronic Work Instruction Software

APPENDIX

References   95

    Industry Standards
    Performance Classes
    Board Types
    Assembly Classes
    IPC Standards
    Military Standards
    ISO Standards

The best experiences of our engineering careers and endeavors can become a lasting legacy for future generations of engineers.

FUNDAMENTALS

The Basics of PCB-RE

> Excellence is achieved by the mastery of the fundamentals.
>
> Vince Lombardi

What is PCB-RE?

PCB reverse engineering (PCB-RE), in its most basic practice, is taking a finished PCB product and attempting to recover its original design data, either in schematic form or Gerber artwork format. I say 'attempting' because not all modern PCBs can be fully and successfully reverse-engineered, in part due to several factors:

- Complexity of design, including density and accessibility of the board
- Sophistication of anti-piracy measures in place
- Insufficient information to identify critical components present

While PCB-RE can easily attain 100% success rate for most through-hole PCBs with 2-4 layers, the level of difficulty increases with the use of large-scale micro-BGA devices and multi-layered PCBs comprising eight or more layers. The manual way of doing PCB-RE becomes impractical if not impossible, and even semi-automatic learning-by-clipping method becomes tedious and error-prone.

More sophisticated and efficient means of doing PCB-RE, however, are employed by big corporations with deep pockets. Engineers in these companies generally have at their disposal expensive equipment such as flying probe machines with associated software to digitize, analyze and schematize the data obtained, or powerful X-ray scanning machines that can peel through PCB layers, stitch and store them up for further processing to reproduce the original PCB artworks.

If you're willing to sacrifice one or two faulty boards, there are also destructive methods known as PCB de-layering or de-laminating, mechanical and chemical, which we will explore later. One interesting field that came out of the reverse engineering discipline is chip-off forensics, an advanced digital data extraction and analysis technique which involves physically removing flash memory parts from a target device to acquire raw data by means of specialized equipment.

Fundamentals

Why PCB-RE?

Despite the high cost and challenges involved, companies and individuals engage in PCB-RE for various reasons:

- Re-create the schematic diagram, in part or full, for repair
- Recover the Gerber data for PCB reproduction
- Re-design the board to circumvent obsolete parts

The first instance applies to repair technicians and agencies that carry out repair of PCBs without any proper documentation, probably from customers who are end-users and not the original equipment manufacturer (OEM), who may have limited recourse either because the OEM is no longer in business or supporting its end-of-life products, or charges a high premium for repair.

Similarly, a PCB may go out of production resulting in not enough new or refurbished stock in the market to keep existing system operational for another five years. In such cases, the customer may be forced to reproduce the PCB by sacrificing a few bad boards to re-construct the layered artworks for reproduction purpose.

Obsolescence is one major issue faced by the commercial and military alike. PCB designers do not have the ability to predict whether parts used in their design will go out of production or stock for whatever reason.[1] Consequently, customers may find themselves in a tight spot when a failed component is no longer available and thus unable to replace it, rendering the faulty board useless.

PCB-RE then becomes a viable option to re-construct the schematics to facilitate re-design works, doing away with the use of obsolete components and replacing with parts that are more readily available. This is usually done without firsthand knowledge of the OEM since there might be possible infringements of intellectual property rights. The legal risks though, is negligible if the re-designed PCBs are meant only for internal consumption and not for external sales to make a profit.

From doing it manually on an ad hoc basis to full-scale automation with reproduction in mind, PCB-RE is increasingly becoming an indispensable discipline in the PCB repair and refurbish industry. The fast-changing market and shortening design-to-product cycles will only see higher demands for such services and practices, not-withstanding the idiosyncrasies and stigma that is attached to this peculiar trade which everyone seems to be doing but no one wants to admit it.

---

[1] I remember involving in an upgrade project for the F-5 radar system with an Israeli contractor, in which I was tasked to write test programs for some of the PCBs that just came out of production. The project manager told me that some of the military grade programmable logic devices (PLDs) were already obsolete and most of the stocks were bought up by them, just in case!

Anatomy of a PCB

I'm no PCB designer, but having worked on PCBs for the last thirty years, written dozens of test programs on different ATE platforms[2] and repaired countless number of boards and electronic modules,[3] I can confidently say that I've seen and worked on enough types of PCBs to be able to adequately take on this subject (from my engineering perspective, of course!).

Nowadays, PCBs are so complex and varied in their design that it is difficult to classify them clearly. Each type of PCB has its own unique form fit and function, so the easiest way to group them is by the products and purposes for which they were built:

- General. These PCBs are usually found in consumer products and computer systems for non-industrial use. Normally they are not conformal coated and operate within the 0-70°C temperature range.

- Dedicated Service. These PCBs are mostly found in communications equipment, business machines and instruments where uninterrupted service is desirable but not critical. The components and materials used are of industrial grade for prolonged operation in temperatures ranging between -40 to 85°C.

- High Performance. These PCBs are deployed in mission critical systems such as life support and military installations, under harsh environmental conditions and extreme temperatures of -55 to 125°C.

My course of work in the air force's E-2C repair bay as a technical ground crew, and later in the home-grown defense industry as a test engineer, offered me the opportunity to handle PCBs of the following types:

- Standard rigid PCBs that were single and double-sided as well as multi-layered.
- Flexprint and hybrid PCBs which were combinations of rigid and flexible type.
- Discrete wire-wrapped boards[4] that used electrical wires instead of PCB tracks.
- Ceramic-base PCBs with dielectric insulation between layers.

Understandably, each type of PCB exhibits characteristics and properties unique to their uses and purposes, whether for prototyping, mechanical strength, durability, good heat-dissipation, or space savings and constraints.

---

[2] You can refer to my engineering bio under *About the Author* at the back of the book.

[3] I'm also a certified PACE technician for that matter!

[4] Israeli defence companies like Elbit and IAI seem to favour this kind of PCBs for their ground repair stations, simply because they are produced in limited quantities and do not justify efforts to go the design route. These are usually VME-size vero-boards with either long component socket pins that allow for wire-wrapping, else they contain sharp blades in which the wire ends are spliced when pressed in to make contact. Such systems usually last between 5-10 years and anything beyond that is really a bonus.

Fundamentals

In its most basic form, a PCB is comprised of the following elements:

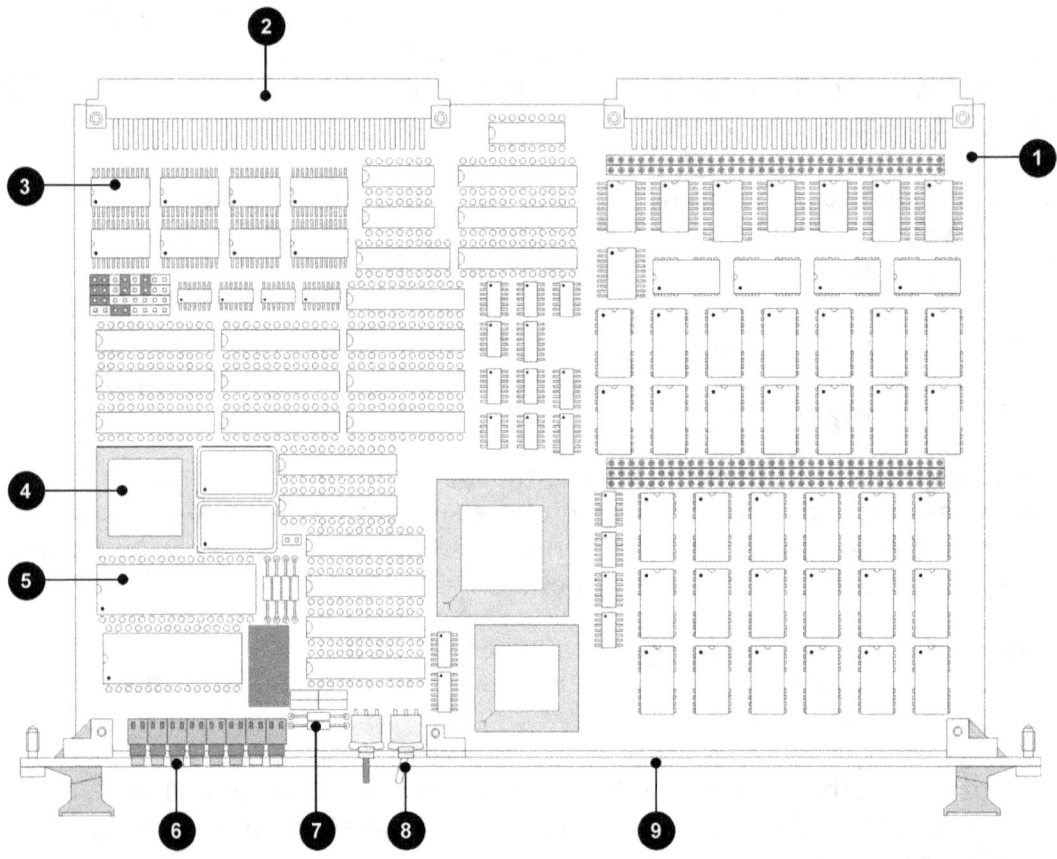

1. Base
2. Connector
3. IC, SMD
4. IC, PGA/BGA
5. IC, Through-Hole
6. LED Indicator
7. Discrete Component
8. Switch/Jumper
9. Mechanical Parts

The base material that make up the body of a PCB is an insulating sheet which combines specific electrical, mechanical and thermal properties in order to meet safety requirements. FR-4 is the most common insulating material used and comprises sheets of woven fiberglass bonded by flame retardant epoxy resin. Certain layers form the core while others reinforce the PCB and are known as the pre-impregnated (prepreg) layers. This is the real estate populated by all kinds of components and parts that make a PCB function as it is designed.

While there are still single or double layer boards around for simple circuit designs, today's PCBs are usually multi-layered due to the power and signal requirements imposed by the components found on them. Routing constraints necessitate PCB build complexity, such as additional layers and the use of blind or buried vias to transfer trace routes from one layer onto another to ensure 100% routing completion.

The Basics of PCB-RE

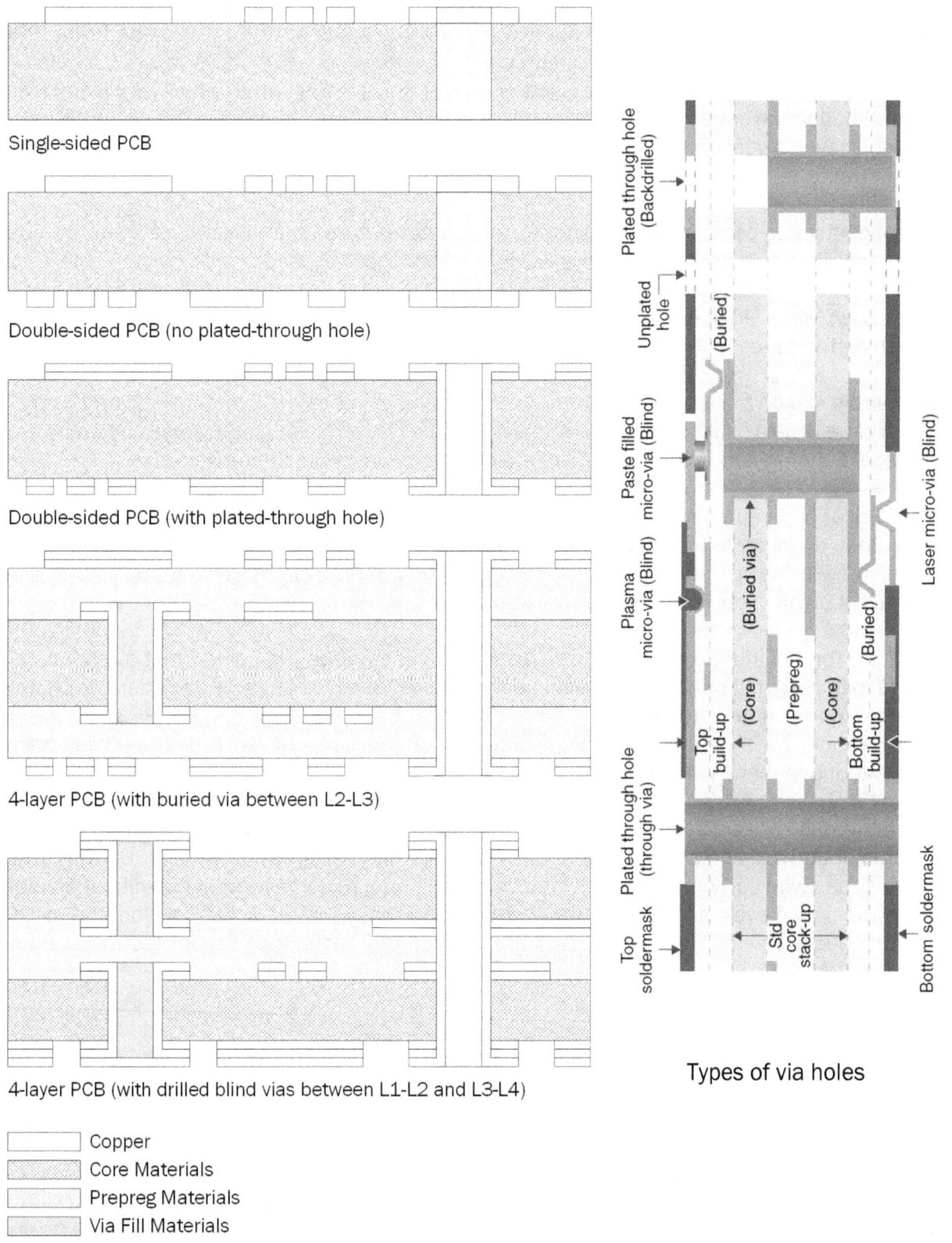

Types of via holes

PCB-RE: Tools & Techniques

Fundamentals

The preceding figures show some typical PCB layer stack-ups. Manufacturers normally use the term 'layers' to refer to the conductive copper foil layers of a PCB, whereas in the PCB layout sense of the word, it includes both conductive and non-conductive layers such as the silkscreen and solder mask layers. Based on this understanding, an 8-layer PCB will have the top and bottom layers of visible copper traces and six internal copper layers, all sandwiched between seven layers of insulating materials which form the core and prepreg that hold these copper traces and their vias (through, blind and buried) together as a whole.

Components also come in various shapes and sizes and are generally grouped into discrete and integrated circuits by their physical forms, and passive and active by their electrical properties.[5] Some are primitive in function while others are an order of magnitude greater in complexity; when engineered into a congruent whole, each component contributes to the overall performance that makes a PCB execute its intended tasks within a system.

It is for this reason that connectors provide the means to interact with other PCBs and the real world, usually via a backplane, and indicators provide visual cues to the statuses and activities of the PCB during normal operation, or flag for attention when a failure happens.

Mechanical parts can also be found on a PCB, either to provide structural support if it has a large body or solder points to dissipate heat quickly and efficiently if it is tightly confined or encased as in the case of metal-clad PCBs, or to hold it firmly in place to reduce vibrational stress in a highly mobile or unstable platform.

The push for miniaturization has seen drastic reduction not only in component sizes but also careful planning and optimum utilization of smaller PCB real estates by designers. What this means is components are not only mandatory to be mounted on both sides, PCBs must also become modular in nature to allow piggy-back versatile configurations, like the mezzanine design found in many VME circuit boards.

Given the rapid advances in PCB designs, the manual approach in doing PCB-RE may not be a viable option, except for partial or small-scale jobs and perhaps limited to PCBs with small board sizes. Even so, component miniaturization will require some form of magnification tools to aid the naked eyes, as well as fine tip probes and a pair of steady hands. Nonetheless, to master a niche skillset like PCB-RE, we will still need to understand the process—beginning with the basics.

---

[5] Most books on basic electronics will adequately cover these components in detail, but if you just want to know how to identify them for PCB-RE purposes, you may want to get a copy of my book, *The Art of PCB Reverse Engineering* (see back pages).

Understanding the PCB-RE Process

Let me get the facts straight: there is no single or standard process to begin with. The process will vary depending on the reason for doing PCB-RE. If the intention is to reproduce the PCB exactly as its original design, that's cloning. If the purpose is to re-create its schematic diagram either for repair or re-design, then it's rightly reverse engineering in the real sense of the term. Cloning, or replicating, is of course illegal as it directly infringes on the intellectual property rights (IPR) of the designer; reverse engineering to reproduce a PCB's functionality instead of the actual design, however, is not.

Now that we cleared the air, let's look at the processes for these two options:

1. Cloning a PCB

    The advantage of working directly on the PCB layers eliminates the need to re-create its schematic diagram—an intensive and time-consuming effort that is potentially error-prone even for a seasoned draftsman. After all, the aim is to obtain the layered artworks to recover the Gerber data and reproduce the PCB bare board. Many PCB design tools do have the ability to translate this data back into schematics but again the process is just as tedious and involved.

    The steps involved in de-layering a PCB are:[6]

    1. Disassembling the PCB and removing components from board surface
    2. De-layering
    3. Imaging and processing
    4. Gerber data extraction

    Extracting the artwork of each layer of a PCB may seem a straightforward and easy thing to do at first sight, but in truth it requires the person involved to possess a few skillsets, unless of course the work is distributed and carried out by different people who are trained in their own areas of expertise. This includes:

    - A good working knowledge of PCB structure and its basic construction to properly and safely de-construct it.
    - The know-how to operate one or more of the following equipment:[7]
        a. an X-ray machine
        b. an abrasive blasting or CNC milling machine
        c. a surface grinder or Dremel tool
        d. a laser cutter
        e. proper handling and disposal of hazardous chemicals

---

[6] Non-intrusive method requires only steps 3 and 4. Imaging includes CT and high-resolution scanning.

[7] Except for the X-ray machine, the other methods are of an intrusive or destructive nature. The choice of use will depend on the availability of PCBs that can be sacrificed.

Fundamentals

- Proficient use of hardware and software to scan, stitch, image-process and map the PCB layers.
- Ability to convert the artwork images into relevant CAE/CAD format to extract the Gerber data and ensure its integrity and accuracy.

When you look at the intricate processes involved, PCB deconstruction and reproduction isn't for the impatient or the occasional apprentice of this art. To attempt it on one's own is almost impossible, even for a moderately complex PCB, without the help of proper tools and equipment. The failure rate can be especially high if you do not possess the required skillsets and relevant experiences.

That said, I know of a company who provides such PCB-RE services. In fact, I visited their premises, saw their high-resolution flatbed scanner equipment and talked with one of the engineers who has years of experience de-layering PCBs. And you know what? This guy uses different grades of sandpaper to do the job! Now I'm not suggesting that this is the best approach even though it is a workable solution. Given the availability and increasing accessibility of better tools and equipment to help cut down time and effort, it would seem illogical if not foolhardy to rely on primitive ways for such laborious tasks.

While the cost of a commercial X-ray machine is out of reach to most people, it might interest you to know that, as recent as the summer of 2015, there was a guy by the name John McMaster who assembled a small-scale, home-brewed X-ray machine to do reverse engineering on ICs and PCBs who was able to get seemingly impressive results!

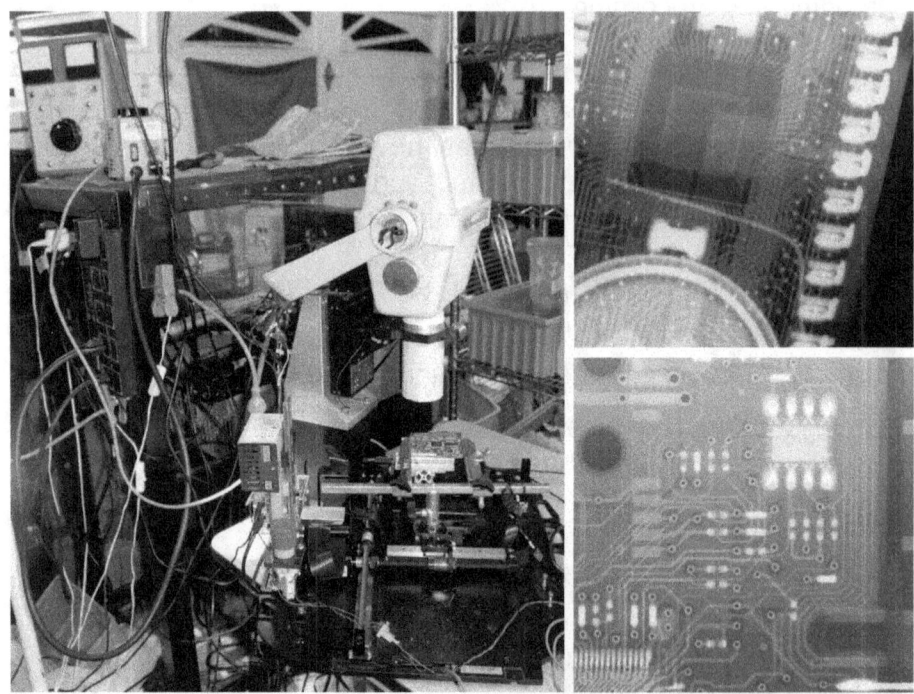

A small-scale, home-brewed X-ray machine (Courtesy of John McMaster)

Mechanical means of de-layering a PCB abound, ranging from handheld abrasive blasting that tests the manual operative skill,[8] to precision CNC milling or laser cutting techniques which afford pin-point accuracy and refined controls. Still, it is important to ascertain the material[9] of the PCB coming under the knife, because once the process begins there's no turning back.

Chemical means are used for solder mask removal and dielectric etching but proper handling and disposal procedures, and ventilation of the workspace must be strictly adhered to due to their hazardous and toxic nature, which can result in fire, explosion, severe burns and other health-related issues.

Non-intrusive or destructive, de-layering a PCB requires you to be well-equipped and take necessary precautions to avoid damaging the board or injuring yourself in the process. Suffice to say, there will always be demand for such services so long as obsolescence and end-of-life for product support issues continue to exist. This topic will be addressed more thoroughly in a later chapter.

2. Reverse Engineering a PCB (PCB-RE)

In a strict sense, reverse engineering when applied to a PCB is defined as the process of recovering the original design, both in the electrical interconnectivity of the board as well as any firmware found in its components. Whereas cloning or de-layering starts with the PCB artwork to obtain its Gerber data, with the possibility of translation into schematics as an option, reverse engineering (RE) a PCB works at re-creating the schematics upfront by discovering how its components are interconnected. There are three approaches to carrying out this task:

- Manual process

The manual approach is most widely practiced and well suited for PCBs of simple to low moderate complexities, or where partial PCB-RE work on a small area of interest is concerned. The advantage of this approach is seemingly low cost if you do not factor in the time and effort; an engineer armed with a digital multimeter with good diode check function and normal eyesight is all that's required to get started.

---

[8] We have a Vaniman Problast 2 micro-abrasive sand blaster that uses aluminium oxide beads to remove tough PCB conformal coatings which cannot be treated in PA93 solution. However, on our first attempt, because we did not set the pressure correctly and control the distance and duration of application, it effortlessly removed more than just the conformal coating!

[9] Not all PCBs are created equal. Some are made of tougher fibre that can withstand abuses (heat, pressure, or corrosion), while some buckle at the slightest touch. I had a female engineer colleague who joked about having to mend a PCB that had so many broken links from doing rework and soldering, she likened the wires that run along the solder side of the board as highways for fast electron traffics.

Fundamentals

The down sides of this process are grave, because of its time-consuming and laborious nature which often result in errors due to human fatigue or carelessness. The onus is on the engineer performing the task to ensure strict adherence to a systematic and organized process in initial preparation work, adopt the right strategy and engage in continuous adjustments to the schematics as it begin to form and take shape.

Steps involved in the manual approach:[10]

1. Preparation work
2. Conformal coating removal (if present)
3. Connectivity verification
4. Schematic diagram creation
5. PCB layout and Gerber data generation (if required)

Preparation work includes: assessing board accessibility, creating a bill of materials, gathering relevant datasheets, and producing a layout diagram of the PCB (drawing or photo of both sides).

Steps 3 and 4 can be performed in parallel if you do not intend to create a netlist text file but work directly with a schematic design tool or a general drawing tool.[11] However, this requires some measure of artistic ability to visualize the placement of various component symbols and organize how their wirings are run and grouped as the work progresses. Alternatively, if you are using an EDA tool, you can perform step 3 alone to produce a netlist, and then import it into the schematic design tool to have it auto-connect and check the electrical integrity for you. This is for engineers who have experience with using such software, which will naturally lead to step 5 if there is a requirement to reproduce the PCB layout and artwork for reproduction.

- Automated process

The automated approach entails the use of powerful but expensive equipment to do the manual and time-intensive tasks of finding and keeping track of connection points. Advances in technologies now allow flying probe tester (FPT) machines to probe PCBs containing very small and fine-pitch components with pinpoint soft-landing accuracy and speed, as well as precise measurement capability for reliable results. Some models can even probe both sides of a PCB simultaneously with such ease and precision that they make light work of even the most complex boards.

---

[10] Steps 1-4 are covered in detail in my book, *The Art of PCB Reverse Engineering: Unravelling the Beauty of the Original Design*.

[11] An electronic design automation (EDA) tool with schematic entry and PCB layout capabilities usually has a steeper learning curve than a general drawing package like Microsoft Visio. If PCB layout is not required, I would suggest using the latter, which not only costs less, is more versatile and a lot easier to use.

All these advantages come at a price, of course; a basic FPT machine costs between 100-150k USD for just a probing unit, and those capable of double-sided probing command even more. And that's not including the cost of software that supports reverse engineering plus the regular maintenance cost of the machine!

Steps involved in the automated approach:
1. Digitize PCB to produce X-Y data of test points[12]
2. Remove conformal coating (if present)[13]
3. Verify and align probe points
4. Commence probing and netlist extraction
5. Export to EDA software for schematic diagram creation
6. Layout PCB and generate Gerber data (if required)

To enable the machine to know where to probe, the PCB must first be photographed with a high-resolution camera to obtain a linearized image, before it is software digitized to produce the X-Y location of each probe point. Depending on the features available on the FPT machine, the process of digitization can involve:

- Scanning the entire board with an integrated CCD camera to obtain a linearized image, then defining each probe point directly on the PCB image with a digitizer software to mark each landing on a physical pad, test point, or component lead,[14] or

- Producing a board outline with matching component layout symbols[15] that overlay the PCB's photo. Once the board layout is completed, the corresponding probe points of each component's solder pads can be automatically determined without the need for visually marking the contact points.

In order for the FPT machine to make sense of these acquired X-Y coordinates data, they must be correlated to the reference designations and pin positions of their component parts. This is taken care of by the bill of materials (BOM) which is usually gathered and input into the system when creating a new PCB-RE project.

---

[12] If BGA components are present, these must be removed to facilitate access for probing. The availability of a bare board would be ideal, else components and fittings that obscure probe points must be moved out of the way. This is the disadvantage of the flying prober approach. The alternative is using JTAG i.e. boundary-scan but this method has its considerations and challenges as well.

[13] If there is conformal coating on the PCB, it is advisable to be stripped clean to prevent residue accumulating on probe tips and cause poor contact problem later.

[14] I know a sub-contractor who builds in-circuit test fixtures employs this similar technique by means of a digitizer probe. But instead of an FPT, he uses a software-controlled CNC machine. The X-Y coordinates he acquired are then used to drill holes on the fixture bed where the spring-loaded probes are to be inserted.

[15] These are standard layout symbols based on the component manufacturers' specifications and should match closely with the profiles of the physical parts on the linearized PCB photo.

Fundamentals

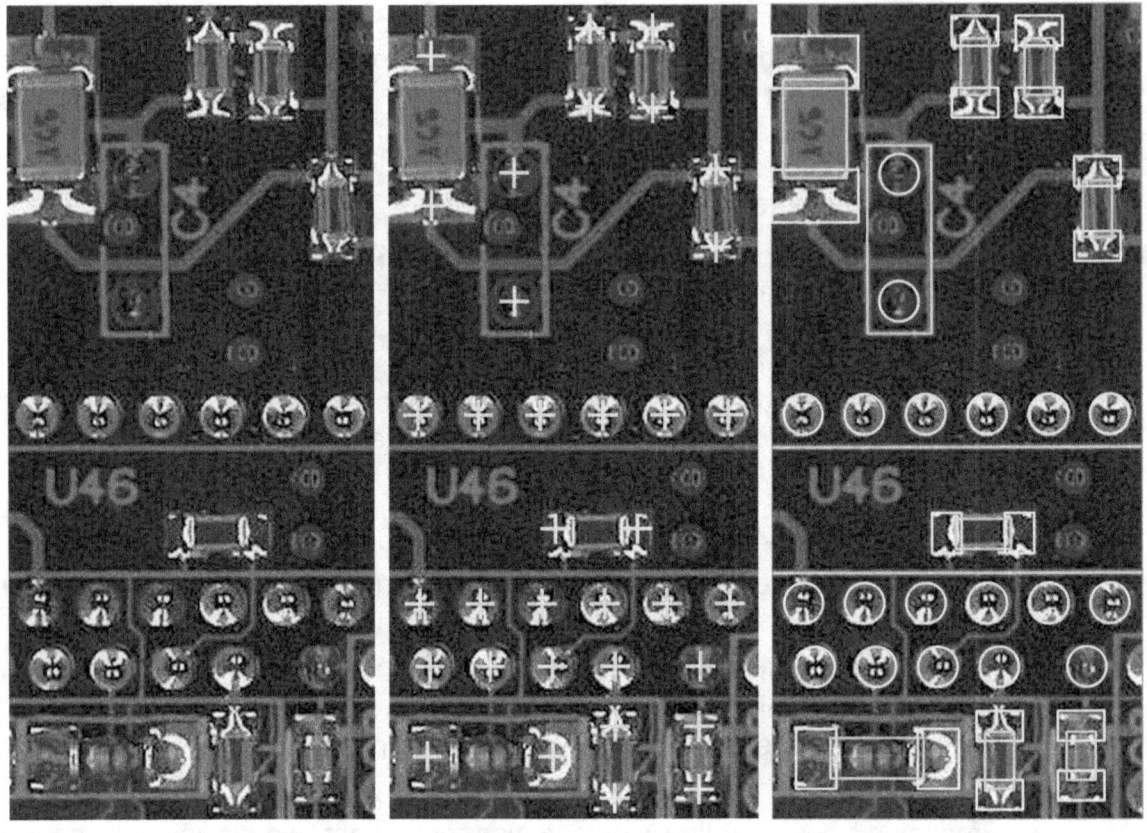

| PCB photo | Digitized (by markings) | Digitized (by symbols) |

Before auto probing can commence, the probe points must be verified and fine-tuned for non-linearity in the original photo image caused by irregularity of components or solder. The power of the flying prober is realized once automatic probing begins. What may take weeks or months to produce manually by hand now gets done in a few hours or even less, depending on the PCB's complexity. The best part is that, besides checking connectivity, the machine can also acquire real values of discrete components like resistors, capacitors and inductors based on component references provided during the digitizing phase.[16] A bill of materials can thus be generated with these measured values filled in.

Once the netlist is obtained and converted to the appropriate CAD data format, either the schematic diagram can be re-created or if only the PCB layout is desired, manual or auto-routing can be employed to re-design the artwork and generate the Gerber files for PCB reproduction.

---

[16] The flying prober machine is also an automatic test equipment (ATE) very much like an in-circuit tester (ICT) except that there is no requirement for a bed-of-nails (BON) fixture to test a PCB. I had worked on several functional and in-circuit platforms and can tell you that test fixtures don't come cheap. But while flying probers can perform basic analogue measurements, it cannot carry out full digital functional tests on ICs with large pin counts, for obvious reasons.

- Semi-automated process

The semi-automated approach would seem to fit the bill for most small companies with limited budgets and manpower constraints, offering a compromise between cost and savings in time and effort. Clip-and-test benchtop equipment has been around for quite a while; with the progression of time and technology, however, the speed and reliability of testing components on-board PCBs has greatly improved. Fortunately, the cost of such equipment has also remained fairly constant and affordable.[17] As demands for PCB-RE tools increase, these equipment manufacturers entered the market with their version of clip-and-learn tools, which is really just an extended function of their existing test tools with the necessary accessories and supporting software.

Steps involved in the semi-automated approach:
1. Create PCB layout or import PCB photo[18]
2. Define components and their placements
3. Remove conformal coating (if present)
4. Clip-and-learn as prompted to generate netlist
5. Export to EDA software for schematic diagram creation
6. Layout PCB and generate Gerber data (if required)

In terms of usage, after defining the components and placement the RE software will guide the operator to clip or probe in clusters, depending on the available channels, and then learn the connectivity of the PCB. The netlist generated can then be exported to an EDA software to re-generate the schematic diagram and to re-design the PCB layout if required. Notice that steps 5 and 6 are similar to the automated approach.

While it sounds simple enough to clip and probe circuit clusters, my experience with one such equipment taught me the challenge is in ensuring good electrical contact between the test clips and the component pins. This implies that the PCB must be stripped clean of conformal coating and residues, the test clips must also be free of oxidation and the grip must be good and not weakened due to mechanical fatigue from frequent use. If the PCB is surface-mounted or mixed type, you'll need an assortment of SMT test clips to do the job as well, and these QFP and PLCC test clips are certainly not cheap!

---

[17] A basic configuration with 256 channels costs around $5-7K but will require lots of manual clipping to be performed. More powerful systems that support from 1024 to 2048 channels can reduce the amount of clipping and speed up connectivity checking but comes with a higher price tag, of course!

[18] X-Y location data is not necessary since only manual clipping is required.

Fundamentals

Evaluating Your Chances

What are the chances of success in doing PCB-RE? It depends on several factors and your approach. Within each approach, the time and effort required will also increase with the complexity of the PCB, while the rate of success and the quality produced will be affected by the capability of the equipment or tool used, as well as the experience of the engineer doing the PCB-RE work. Quantifying the methods and their related factors into a single chart offers a rough but straightforward comparison as shown in the table below:[19]

Methodology	Time	Effort	Cost	Success	Quality
Manual	4	4	1	3	3
Flying Probes (FPT)	2	1	3	2	2
Boundary Scan (JTAG)	2	3	2	3	3
Clip-and-Learn	3	3	3	2	2
Delayering[20]	2	2	2	1	1
X-ray CT Imaging	3	3	4	3	2

Weightage: 1 equals least time and effort, lowest cost, best success rate and highest quality.

Another way of representation would be individual SWOT charts that give a sweeping view of each approach's strengths and weaknesses, as shown overleaf. The closer to the center (i.e. bullseye) a factor is, the better the weightage and vice versa.

It should be noted that except for the manual approach, the success rate and quality of the end results rely on both the hardware and the software components involved, whether it is the algorithm that drives the FPT machine or guides the engineer doing the clipping, the image analyzer that maps, stitches, and processes the layers, or even the back-end that does the conversion from the raw data into Gerber format or CAE outputs.

No matter what you opt for, it is important that you have a good understanding of the process involved, not just the cost incurred. Other more specialized aspects of PCB-RE such as mobile forensics may require formal training to be proficient in performing the tasks yourself, unless you prefer to engage a professional to do the job. These exotic, CSI-like practices on the chip-level are too niche and involved and will not be covered in this book.

---

[19] Obviously, it's not a one-size fits all comparison chart, since the types of PCB are so vast and varied, but doing a case-by-case comparison would prove tedious if not unfeasible. A 'rough' estimate based on a moderately complex (4-layer) PCB as an entry point is given here.

[20] Delayering can be destructive (mechanical, chemical) or non-destructive (X-ray imaging). For simplicity's sake, I'm referring to the physical aspect of removing individual layers by machines.

The Basics of PCB-RE

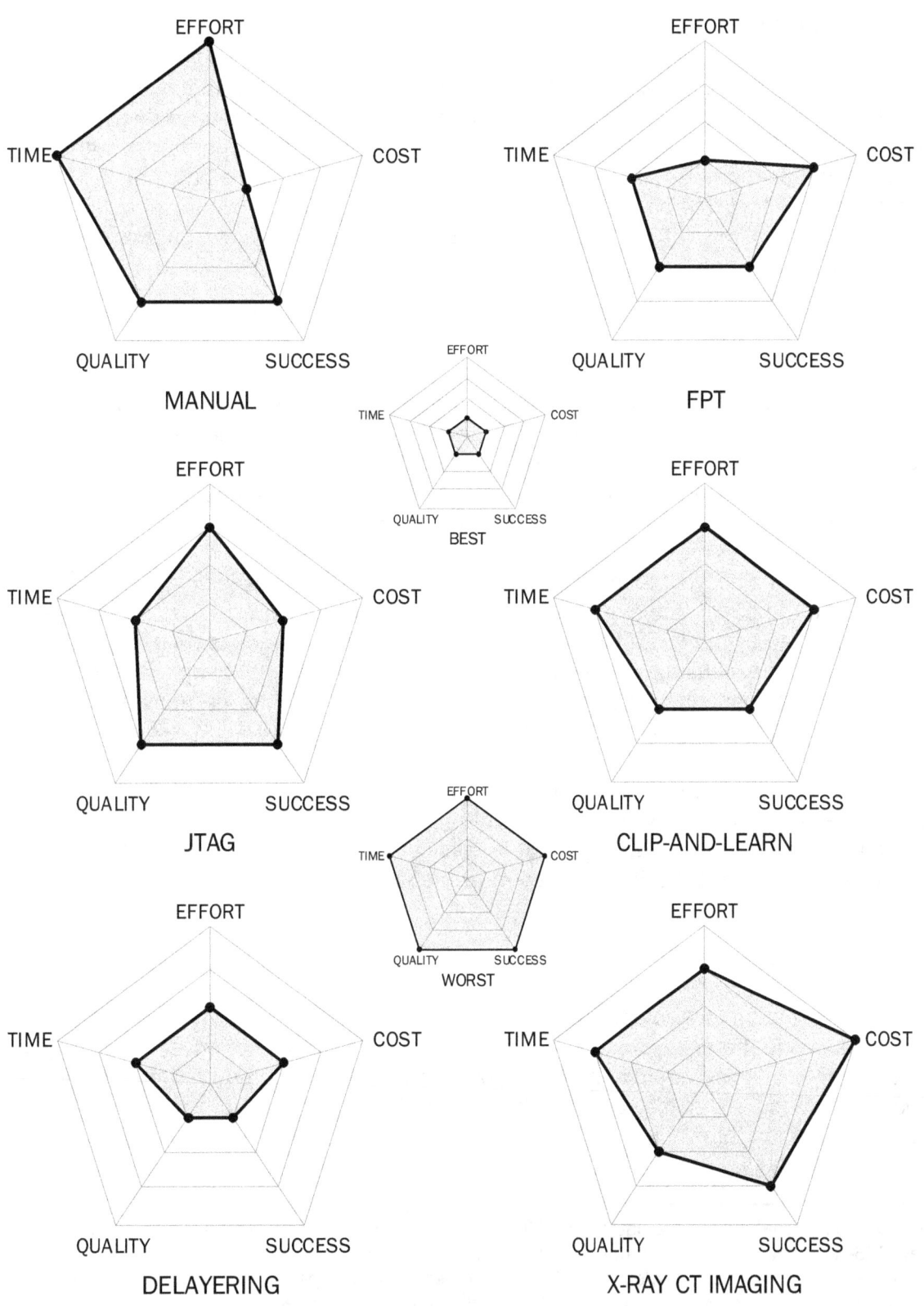

PCB-RE: Tools & Techniques

Fundamentals

Going a Little Further...

The figures and charts given are based on my perception and understanding, from researching as well as personal experience. Those who have worked in the respective fields longer than I did, or knew some intricate details that I don't, may disagree. That's perfectly alright. After all, as I mentioned in footnote 19, it's just a 'rough' estimate to provide a ballpark comparison. In the course of writing this book, I've gained a better appreciation of the various approaches through my interactions with different industry experts, and thought it would be good to share a little more on three of the methods:

FPT

One factor that must not be overlooked is the programming time, which can take up a sizable portion of the preparation work, especially when there isn't a silk screen, or if a BOM needs to be generated, etc. Also, validation of the netlist is a must after learning is completed. In terms of effort, the machine does the hard work but at a cost—FPT machines with PCB-RE capability are expensive.

Success rate is subjective, though. There can be issues with FPT, especially if there are broken barrels inside the board due to gas bubbles from de-soldering devices. These are opens that can't be detected by the FPT. Also, repeated probing of contact points on a PCB can damage the pads over time, causing potential opens as well. Therefore, it is important to have more than one board to work on to allow an independent validation on the learned netlist, preferably with the optical delayering technique, if possible.

As for quality, well... it depends on what you want to achieve. If a schematic is desired for repair, FPTs work great, provided the validation process mentioned above is carried out. If a replacement board is needed, FPT is not ideal since they cannot replicate the geometries of the original PCB—no form, fit and function whatsoever.[21]

X-Ray CT Imaging

X-ray is a fascinating technology but it's not magical or mythical when it comes to PCB-RE. Its usefulness lies in the ability to see beyond the physical through penetration imagery, but as with all image scanning equipment, the higher the image quality required, the longer it takes to acquire, tweak and stitch, along with additional mitigating factors related to the object under examination thrown in.

---

[21] For example, signal timing associated with differential pairs is not recreated in the learned netlist from a FPT. A new PCB can be fabricated, but it may or may not work in a system since it may not handshake properly with the other system components. An optical delayering process produces perfect form, fit and function. All traces, shielding, ground and power planes are recreated exactly as in the original PCB—including conductor thickness, dielectric material type and thickness, etc.

Typically, it takes more effort than delayering but less than FPT. This is because there are several steps in the process to extract data from X-ray images; often, there will be artifacts or blind areas due to beam hardening so manual probing is needed to verify the connectivity. Sometimes it is impossible to get under BGA packages, epoxy sealants, etc.

When it comes to cost, only large companies with deep pockets can afford such a machine. Unfortunately, the success rate can be mediocre or even lower for these expensive systems due to the blind areas and a variety of other issues (e.g. PCBs that are not co-planar). It is difficult for CT scans to follow a single conductor layer in all but the smallest PCBs. Large PCBs with high density and layer count are also a big problem.

Quality wise, CT Scan can provide form, fit and function for inner layers in successful cases. When used in conjunction with optical systems, the X-ray images can be scaled in X and Y directions to match the outer layers of a PCB. X-ray machines are generally low resolution, dimensionally incorrect and susceptible to blind spots. They are getting better, but achieving consistent high-quality images of high-density PCBs can be a challenge. That said, they work well for simple, small boards.

Delayering

If you can and are willing to sacrifice a board or two, the delayering approach offers the best chance of success in recreating all the necessary data needed for rebuilding PCBs, even in instances which the boards suffer certain level of damages—lifted pads, internal gas bubbles, burnt, etc. While it may not be the fastest means of PCB-RE, small simple PCBs can be completed in just hours while large, high-density, boards with 16 layers or more can take a couple of weeks. Thankfully, it does not require a rocket scientist to do the job; in fact, the process can be easily managed by an operator.[22]

Compared to the other PCB-RE methods, delayering is surprisingly affordable. Success rate is also probably the highest, but one of the greatest incentive in this approach is the advantage that recreated boards do not need to go through the tedious design testing and certification phases. Since each replicated board will be an exact copy of the original,[23] it is guaranteed to work when plugged into the system.

In the end, it comes back to the question of what you want to achieve with PCB-RE. This will very much determine which approach(s) best suit your needs and budget.

---

[22] The best operators are young people that are good at video gaming! Of course, having a supervisor that is PCB literate, preferably an engineer, to oversee the process is important but an operator who enjoys playing video games will do the job adequately. Now, who says you can't have fun during work?

[23] In fact, the characteristics of the original board (signal integrity, noise, etc.) is faithfully reproduced. Of course, this is provided all the components can be sourced and assembled on the replicated PCBs.

Fundamentals

A Rare Interview

It's not often you get the chance to talk to someone in the PCB-RE industry with a wide portfolio in terms of exposure and experience, much less to share what they know about their tools and trades. So when I was fortunate enough to get in contact with William Loving, the CEO of ScanCAD International Inc., I was tempted to ask him a few questions:

Me : Why is PCB-RE fast becoming an important and indispensable industry?

Bill : PCB-RE is a respected and much needed business activity in the world today. The tools and processes covered in this book are used globally by many companies and even governments to keep legacy systems running and to recover design data that has been lost.

Me : Do you think PCB-RE tools are susceptible to abuse for piracy purpose?

Bill : Any tool or process can be used for good or bad——it's a function of the user. We like to think that our family of customers are working for the good of all concerned. Sure, there will be some bad apples in the group, but we think it's very, very rare in our crowd. Like your good self, we do not condone illegal or unethical activity that will hurt the interest of any individual, company or organization.

Me : How do you think ScanCAD can benefit the PCB-RE community or companies?

Bill : Having been in the business for 27 years, I see PCB-RE as having a greater impact and positive influence not just on a community or company. Having worked with over 1000 companies across 48 countries, I dare say that our products contribute to the overall well-being of our environment. How, you ask? By keeping legacy systems running, we are extracting more value from our earth's limited resources that have already been used to produce existing electronics and systems. Keeping them out of landfills and in operation is a good steward thing to do. This helps on several levels: reduces the waste, extends use of resource and productivity, reduces air and water pollution since new systems do not need to be made to replace the old, etc. In some cases, a PCB may represent only a fraction of the mass of a large system, yet this PCB can cause the entire system to be scrapped if it can't be replicated. Now, that's a big multiplier effect!

Me : Do you foresee yourself and your team doing PCB-RE for the next 30 years?

Bill : ScanCAD will be around for a long time for sure. Like my colleague Jeff said, we have a dedicated and very capable team, with very good people who strive to be a positive influence in the world and do what's right. If I might add, they are also fun to work with! PCB-RE is a strange little niche—— one that we enjoy doing——and one that your book has given us some interesting perspective to look at too.

Why PCB-RE? (Again)

Previously, I mentioned three main reasons PCB-RE are carried out by companies and individuals from the perspectives of repair, reproduction and redesign due to obsolescence. The actual spectrum of people involved and their motives (or motivations), however, are much broader. In fact, I can think of six of them:

1. Rivalries and Competitions

    Put it this way: everyone wants a piece of a lucrative pie. The problem is not everyone can invent, but they can certainly innovate and improve on existing products, especially the popular and fast-selling ones. What better way than to study a current working model instead of starting from scratch? The Japanese started this trend with household appliances by studying the ideas behind the products the West created, then making them smaller, more efficient, and more affordable. This was followed by the Taiwanese[24] who for a time ruled the PC hardware market. Today, Chinese board makers employ this same tactic in the digital industries.

2. Curiosities and Education

    Some people are born with a curious disposition that drives them to pry open anything to find out what makes it ticks. There's nothing quite like learning from the experiences of real-world engineers which textbooks and classroom lectures cannot provide. There is a saying, "You can hear a hundred times concerning a subject, but it cannot be compared with an actual encounter." Active learning is the fastest and most effective approach to becoming an expert.

3. Legacies and Continuity

    Remember the days of floppy drives and diskettes? Those were the good old days we look back fondly on but would rather not go back. Similarly, valuable information and resources are stored in old medium and equipment which needs to be migrated to more modern platforms for continual operation. Examples are the early DEC VAX and PDP computer systems which have since been emulated on powerful PCs and seen a rise in performance magnitude as well as ease in data storage and backup.

4. Varieties and Extensions

    It is said that for every gadget invented there are at least three accessories associated. Big dreams always attract small dreams, and sometimes their smaller counterparts can become more successful or even the driving force for continued popularity by extending the usefulness of the original product. So while companies like Apple keep their designs a

---

[24] Together with Singapore, Hong Kong and South Korea, these countries underwent rapid industrialization, technological innovation and development with consistent phenomenon growth rates in the 80s and 90s, they were dubbed the four Asian dragons.

Fundamentals

closely guarded secret to secure their market shares, it could not stop others from coming up with their own ideas and ingenuity to extend the original products' functionalities.

5. Self-Preservation and Survival

A company that outgrows itself may sometimes need to fall back on an old but successful product design, only to realize that the original files or documentation no longer exists or are misplaced and lost. They will have to resort to PCB-RE to recover the valuable assets, or else to further enhance and secure them to prevent others from attempting to reverse and duplicate their products.

6. Villainy and Malice

Like Star Wars, there is a 'dark side' to the practice of PCB-RE. Some do it for fame and notoriety while others for gain or just to mess things up a bit, but the result is always detrimental to their victims and targets. Backdoors, weaknesses and design flaws are often studied and exploited to inflict damages through loss of data or theft of information. So which side are you on, reader?

Houston, we have a problem!

Well, who doesn't anyway? Like any other engineering field, PCB-RE has its own set of unique challenges and difficulties that can make it seem like a black art to the uninitiated.[25] Even an experienced practitioner will occasionally get stumped by state-of-the-art anti-tampering measures put up by the product designer or manufacturer, run out of ideas to circumvent an obstacle, or work confidently on a seemingly promising project only to realize it's a futile effort.

As far as PCB-RE engineers are concerned, every PCB or chip[26] they work on is a new ball game that requires diligence and dedication, but without the sure guarantee it would deliver the results. Feeling flabbergasted already? If you're still game, let's look at what surprises are lined up to frustrate us.

---

[25] My ex-colleagues would pass by my desk, look at my RE work, shook their heads and gave me the kind of look, like you're one crazy guy to even consider this option. But they certainly appreciated its value when I handed them the schematic drawings for their repair works.

[26] With increased integration and sophistication in design, PCB-RE is no longer limited to just board level kind of reverse engineering. In this chapter, both PCB and chip related issues will be treated generally, so it's important to have a broad mindset and keep your options opened at all times.

Anti-Tampering Measures

Many types of anti-piracy measures are invented and employed by product manufacturers to mitigate the risk of design thefts, breach of security and loss of revenue. To be sure, no tamper-proof measure is effective against every kind of hacking; however, they are put in place for a variety of reasons:

- Tamper-resistant

    One of the easiest ways to discourage snooping and dissuade would-be hackers is to limit access to the hardware. This is usually achieved by using specialized one-way screws with lock-tight applied to the threads, enclosing components inside opaque casing or potting compound, and applying epoxy encapsulation on ICs that contain firmware and important data.

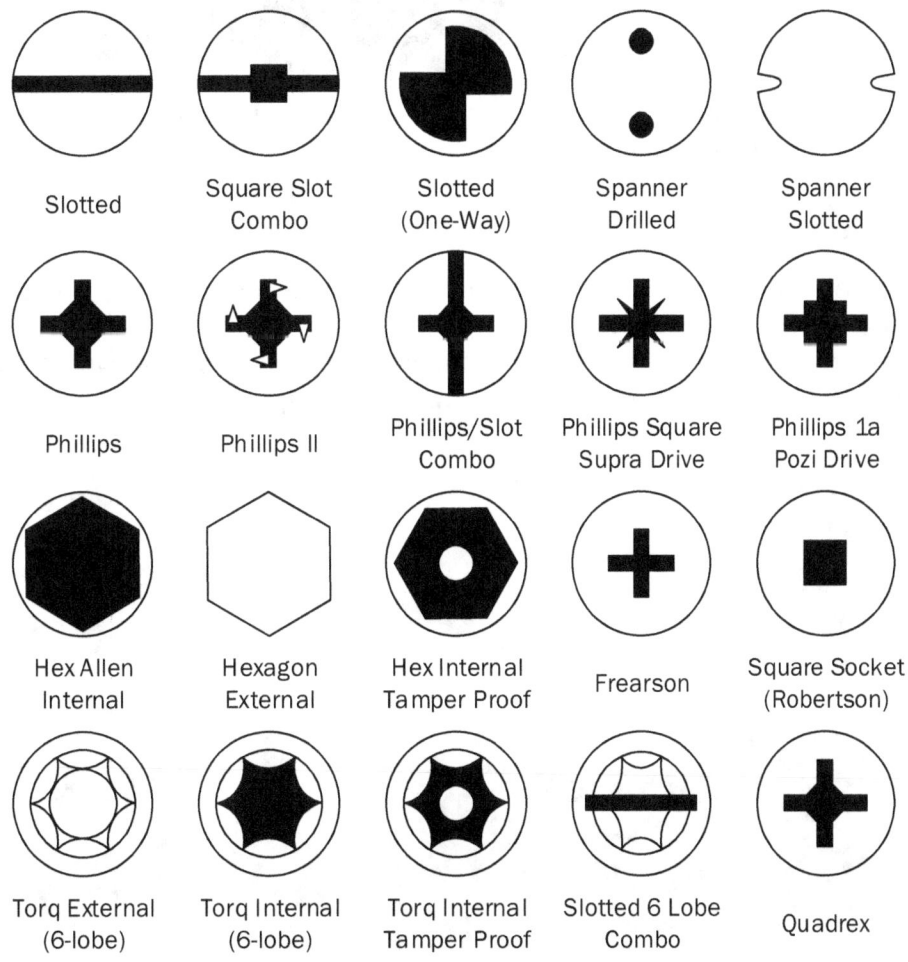

Security bits and one-way screws

Fundamentals

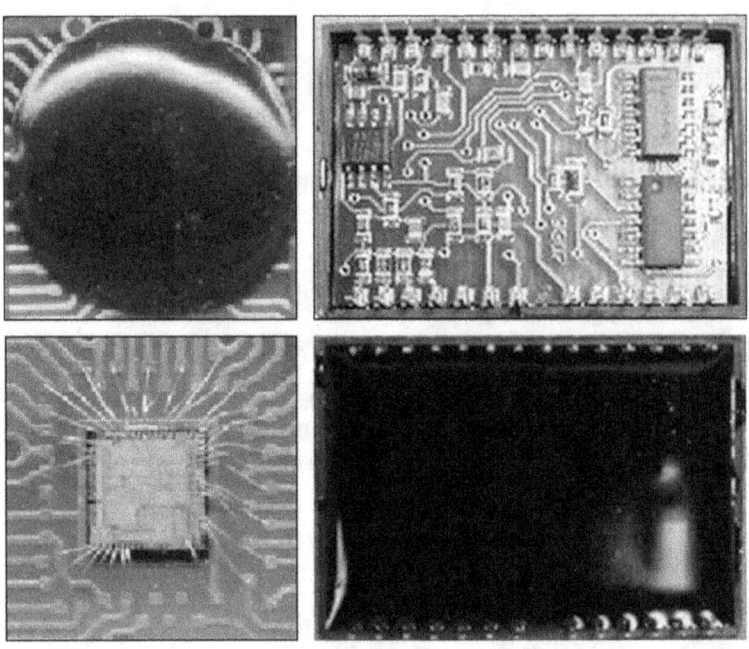

Encapsulation of a COB* (left). Potting of a PCB (right)
* Chip-on-board

- Tamper-evidence

    If access is compromised, the next solution is to create evidence of the violation by means of identifiable breaches, such as tamper-proof decals to void warranty, special wax, seal or glue that breaks upon force of separation, etc. This will cause the average end-users to think twice before voiding their product warranty and support.

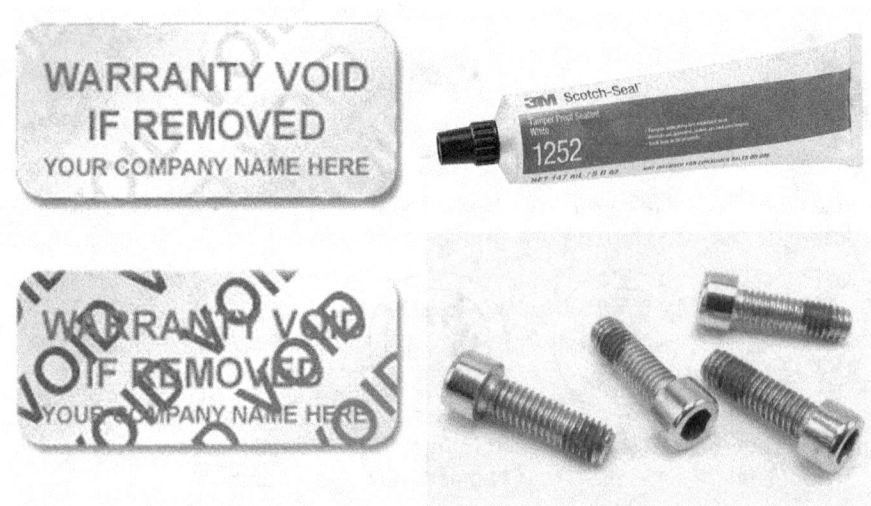

Warranty void decals (left), tamper proof sealant (top-right), and nylon glue processed screws (bottom-right).

- Tamper-detection

    Pressure switches, temperature sensors or puncture detector can be installed within or in hidden locations that alter their positions or characteristics upon violation of the product's integrity.

 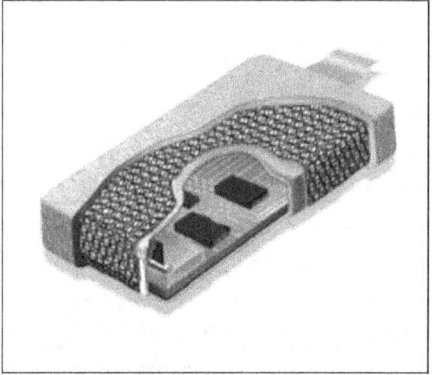

Tamper-respondent sensors (left) and anti-tamper intrusion sensor implemented against physical hacking (right).

- Tamper-reaction

    Such measures go one step further by activating counter-measures upon detection of violation. It can be built-in intelligence that detect tampering from normal operation and perform self-erasure of firmware, usually stored in flash memories, a hardware reset that requires authentication and re-activation from the vendor, or a more drastic self-destruct action that renders the product unusable.

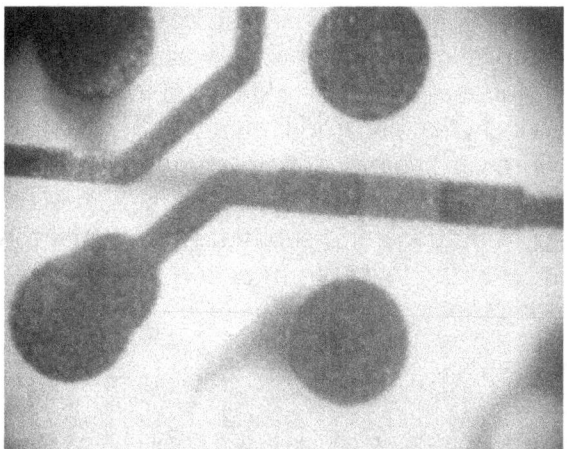

Resistor embedded in a PCB track (Ohmega Technologies, Inc.) Some chips have embedded fuse resistors that incinerate with the application of external voltages.

Fundamentals

The above anti-tampering measures are usually applicable to PCB-level and device protection, at the least to discourage such attempts, or else to make it difficult or prohibitively expensive and time-consuming to do so. The table below lists some of the most common practices by manufacturers implementing anti-RE schemes, and shows the impact on the manufacturing process, cost incurred both to the designer and hacker:[27]

PCB-RE Preventive Measures	Manufacturing Impact	Design Cost	Hacking Cost
Custom screws and unusual fittings	2	3	1
Full potting and partial encapsulation	3	2	2
Silkscreen or component labels omission	2	2	2
Custom ASICs and unmarked ICs	3	2	2
Limiting access using BGA components	4	2	4
Unexposed inner layer signal routing	4	3	3
Intentional multi-layering of PCB	3	4	5
Blind and buried vias	5	3	3
Dynamic signal or bus jumbling	1	2	2
Hidden routing using ASIC	3	5	4
Hidden routing using FPGA	3	3	3
Eliminating JTAG and debugging ports	3	2	2

1: Very Low   2: Low   3: Moderate   4: High   5: Very High

It must be mentioned that anti-RE measures are also applicable and practiced on a chip-level, except that the level of sophistication is of a microscopic proportion and involves more intricate equipment that requires specialized skills and expertise from makers and hackers alike. Common approaches include cell camouflaging of the IC layout, obfuscation of function through interlocked code words, or the more complex means of active metering using physical unclonable functions (PUFs), external key activation using the EPIC[28] technique, and bus scrambling by reverse bit-permutations and substitutions. Again, there is a trade-off between the IC's floor area overhead and the level of protection that the manufacturer must carefully consider before implementing these measures.

---

[27] Referenced from ACM Journal on Emerging Technologies in Computing Systems, Vol. 13, Issue 1, Art. No. 6, Pub. date: December 2016.

[28] EPIC (Ending Piracy of Integrated Circuits) is a novel method that protects chips at the foundry by automatically and uniquely locking each IC. In EPIC, every IC is locked by asymmetric cryptographic techniques that require a specific external key. This key is unique for each chip, cannot be duplicated, and can only be generated by the IP rights holder. (Reference: IEEE Xplore Vol. 43 Issue 10, October 2010, Pages 30-38.)

But is it Good Enough?

> Resistance is futile. You will be assimilated...
> The Borg

Given the relentless efforts and innovative ideas to counter hardware and software hacking, the question is: Does it work? Hackers are not the average engineer that we know but highly intelligent and motivated individuals who make hacking their life's ambition—and nothing, not even tamper-proof measures, will stop them. Reminds me of the wordless comic strip 'Spy vs Spy' published in the MAD magazine featuring two secret agents, one white and one black, each trying to outsmart the other guy. So, the answer is: You win some, you lose some. In other words, there is no absolute full-proof solution against a determined hacker, and even if it manages to deter him now, it's only a matter of time before he or she gains access to the data. On a serious note, though, tamper-proof measures seldom prevent or stop an expert hacker, probably only slow the attempt, that's all.[29]

Let's look at how hackers defeat some of these anti-tampering measures:

- Specialized or one-way screws

    Just as there are skeleton (or master) keys that can easily open a wide range of locks, there are also no lack of universal screwdrivers with changeable tips to adapt to different types of screw heads, and these tools are surprisingly cheap and available online, in flea markets and hardware stores. Manufacturers certainly are aware of this fact and they may resort to some sneaky tricks, if not for any reason, just to add frustrations to those bold enough to tinker with their products.[30]

- Potting and encapsulation

    If you ever wanted to be an archeologist but ended up being an engineer, this may be a good opportunity to experience firsthand what it means to painstakingly and patiently uncover a buried treasure (or fossil), albeit on a smaller scale. Soft potting compounds such as silicone is easily removed with sharp surgical tools. Polyurethane, though tougher than silicone, can be treated with Foam Off-MP[31] to soften before subjecting it under the

---

[29] And there are many of them out there, judging by the numerous discussions posted on various electronic forums on this topic, some bordering on self-brag about their successful exploits or outright disdain or contempt for the ineffectiveness of these measures.

[30] There were a couple of instances during my work when these one-way screws got rounded out because they were so tightly fastened by lock glue or made of soft materials like brass that were meant for one-time fastening. The only way out was to drill through the stuck bits and then re-tap the holes and replace with new fitting screws. It was a lot of work but still it's better than sending the whole unit overseas for repair which would have incurred greater turn-around time and cost—if OEM support is still available, that is.

[31] This is an environmentally sensible solvent from the company GSP Inc. that is bio-degradable, non-toxic, non-corrosive, residue-free, low VOC and non-flammable. A better alternative than acetone or methylene chloride, both of which are highly volatile and toxic to humans.

Fundamentals

surgical knife. Epoxy resin compounds are generally hard and requires more effort to deal with, so it's advisable to X-ray and ascertain the layout of the components to avoid damage. For personal health reasons, it's important to observe safety precaution and put on your PPE[32] when handling chemicals.

- Removing or obscuring IC part number

Small companies without big budgets will usually opt for more cost-effective measures to protect their products from PCB-RE. One of the earliest practices is grinding part numbers off certain ICs so they become anonymous 'black-boxes' to prying eyes. Depending on the resources or tools you have and the types of IC you're dealing with, the method to decipher the identities of these 'unknown' ICs will vary:

a. Without resorting to removing or destroying anything, the first approach is finding the power and ground pins, and if it's a CPU-like chip, the clock and reset pins. Compare these against popular CPU pinouts and you might just find a match.

b. Surprisingly, some designs use only common ICs so the chip in question may just be a TTL or CMOS device. If you have a portable IC tester, simply take out the IC, insert it into the ZIF socket and select the 'unknown' mode, then let the IC tester step through the logical truth-table tests and identify the device for you. Some better IC testers can even test and identify 'unknown' memory devices.

c. If nothing else works, you can always fall back on the decapsulation process. Every IC has a marking on its die that reveals its identity.

Die markings (magnified) of the VIC2 6569 R3 chip.[33]

---

[32] Personal protection equipment.

[33] Source: mail.lipsia.de (Courtesy of Michael Huth)

The Basics of PCB-RE

- Hiding or limiting access

    Whether it's BGA or unexposed (blind or buried vias) multi-layering PCB, today's CT scan technology and equipment can effortlessly peel off each layer to reveal the underlying artworks, and through the process of stitching and digitizing, enables 3D viewing and walk-through of the stacked layers. It's definitely an expensive PCB-RE endeavor for the hacker but so is the cost of manufacturing with such designs in the first place.

- Eliminating JTAG and debugging ports

    These only inconveniences but is seldom effective against the most determined hackers who can figure out the topologies and connectivity of PCBs and their associated components.

- Firmware protection[34]

    Security bits and locks are usually found in memory and microcontroller chips to prevent firmware access and readouts. Sometimes, non-invasive attacks such as varying voltages or temperatures can circumvent these measures.[35] For instance, the PIC16C84's security bit can be easily cleared by repeated write accesses with the VCC voltage raised to VPP−0.5V; the Dallas DS5000 processor's security lock can be released by subjecting it to a simple short voltage drop, if applied correctly. If non-invasive attacks do not work, then it's time to get physical. The first step is decapsulation to expose the die,[36] then the lock bit cell of the on-chip EPROM is located and erased using a focused UV light. If a permanent lock bit is used, then a laser cutter microscope is used to sever the security link.

- Embedded routing in ASIC or FPGA

    A more sophisticated means is to make a programmable or custom-designed IC part of the PCB signal routing—in other words, passing the tracks through a specialized chip. This approach will most certainly deter many PCB-RE practitioners who have no experience nor expertise in chip-level analysis. However, those with the skills and equipment can still reverse engineer even an ASIC by sectioning the chip with the help of a scanning electron microscope (SEM). It's a costly affair for both parties, of course!

These are just some ways to counter anti-tampering measures and they are by no means exhaustive or 100% effective. Every attempt on a new prospect will be a fresh challenge, but that's what makes hacking fun and engaging in the first place, isn't it?

---

[34] Examples provided herein are cursory for interest and information, and do not constitute ScanCAD's or the author's approval on piracy or violation of IPR. The actual processes may be more complex and involved.

[35] Some security processors employ sensors that detect voltage or environment anomalies and reset their content, or disable normal operation if the clock frequency goes too low, to inhibit single-stepping attacks. But these measures can backfire and cause operational problems that frustrate and turn away real customers.

[36] The top layer of the IC is usually removed mechanically or with fuming nitric acid to reveal the wafer and bond wires. Commercial IC package removal equipment are available that do a cleaner job than by hand.

Fundamentals

A Little Interesting Anecdote

To wrap things up a bit, let me share with you an experience I had with a semi-automated benchtop PCB-RE system:

Back in 1998, my department landed on a test program development project involving several PCBs from a mobile radar system. That's when we decided to procure a pinpoint test system which, at the time, was the only one that offered the clip-and-learn RE feature. My manager knew I was capable of PCB-RE but he thought it would take far too long to do it all by myself and could possibly take a toll on me. So he assigned a second engineer to learn the RE system to assist in generating the schematics for the rest of the test program developers.

I was curious how this semi-automated system measured up to its claim, so I requested to have my board, a Radar Processor Module (RPM), be the first to put it to the test. It's a huge card comprising many socketed TTL chips and draws about 7A when powered on. Given that these are all dual-in-line ICs, it should not pose too much of a problem for the equipment to handle, or so I thought.

Much to my surprise and chagrin, after five days of laboring on it, this engineer passed me a stack of A4 papers with what seemed like maze puzzles of crisscrossed wires and awkwardly positioned component symbols. "Is that the best it can do?" I asked in disbelief. She looked at me apologetically, managed a smile and shrugged her shoulders.

I reckoned that I was better off doing it myself instead of relying on some machine that could not even produce a decent schematic diagram. The only consolation I had was the generated netlist for reference to cut down on tracing the connectivity. Overall, it took me about a week and a half to trace, verify and draw using Visio Technical 4.3. Meantime, I found some errors in the netlist which, I suspected, were caused by poor clip contacts or missteps during the learning process.

As it turned out, the schematics spanned five A3-size pages.[37] Together with the PCB layout diagram, I triumphantly showed the engineer my masterpiece and grinned, "That's how a good schematic diagram should look like!"

---

[37] As it is, I can only provide thumbnail views of my work due to the sensitivity of the content. But it should give you a good idea what Microsoft Visio is capable of and dispel any doubt readers might have about my manual PCB-RE ability (grin!).

The Basics of PCB-RE

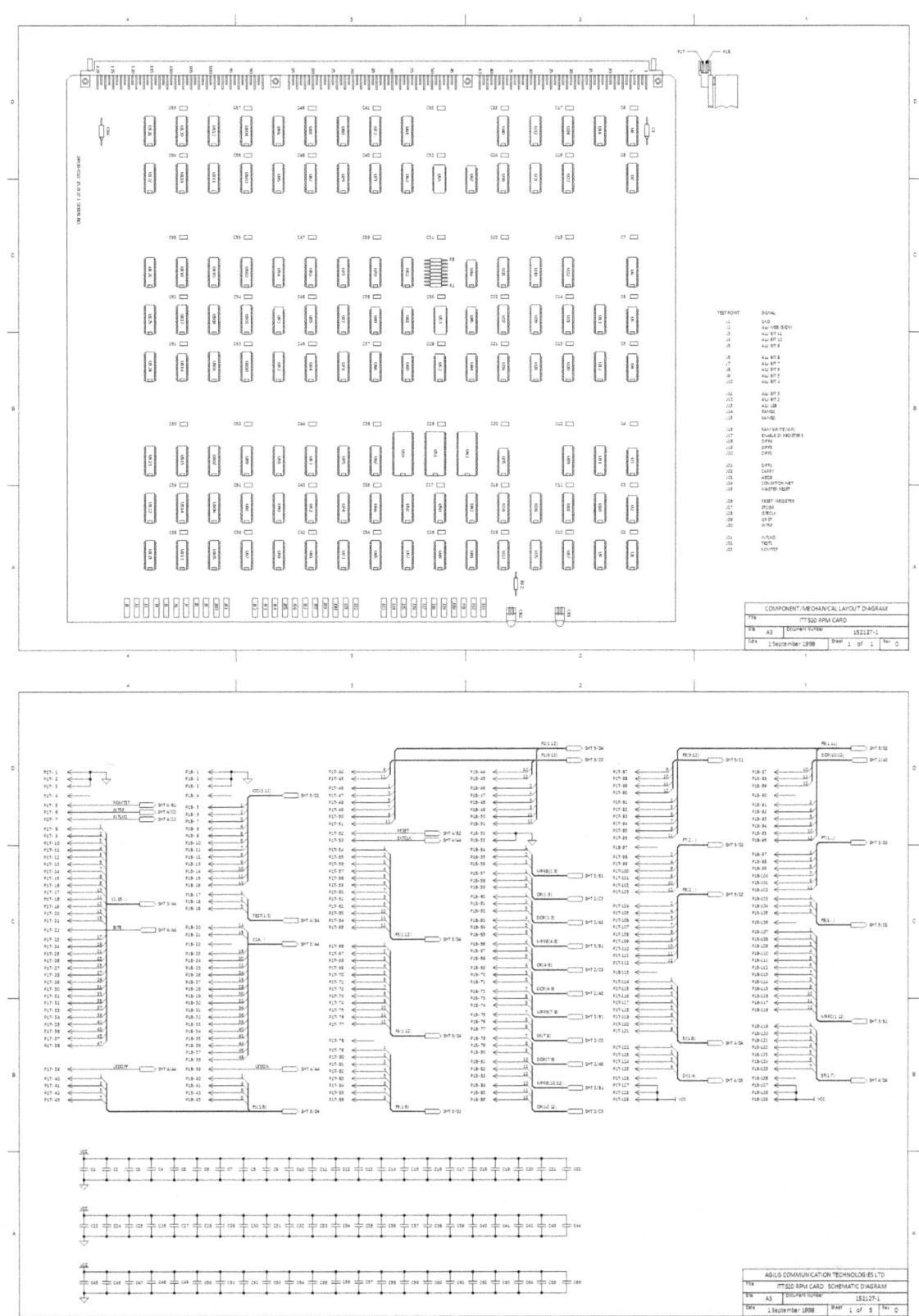

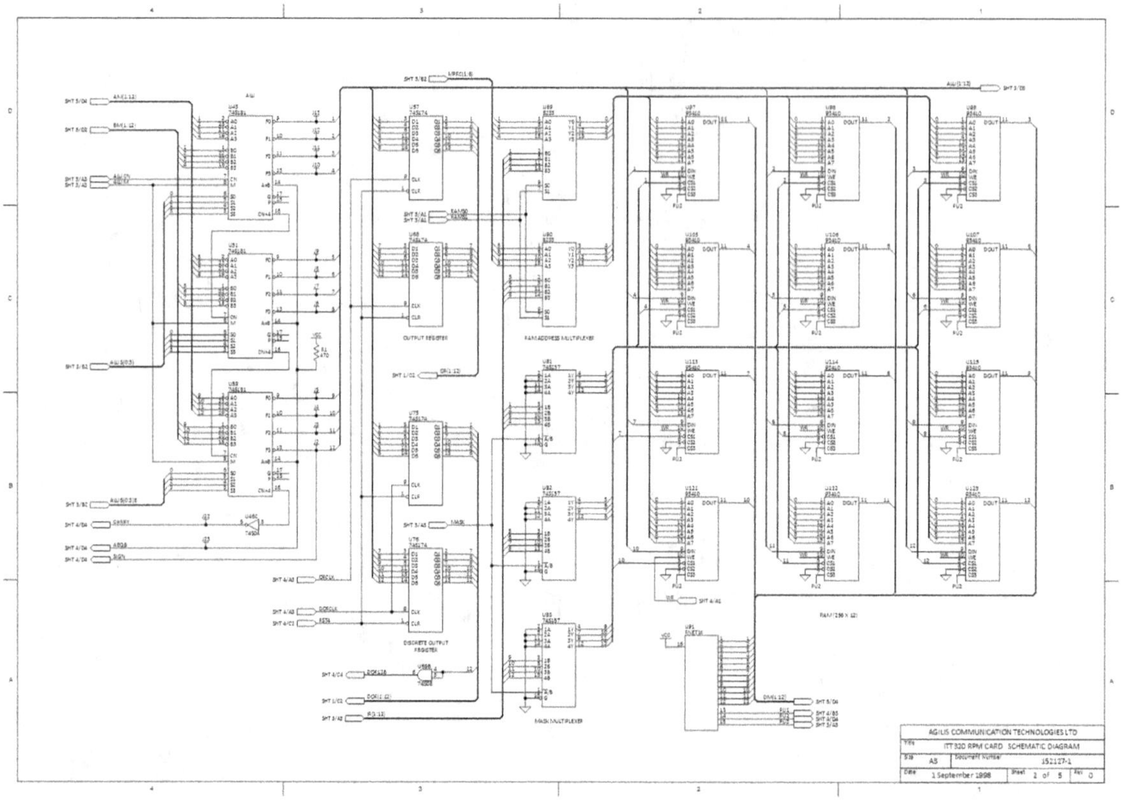

Fundamentals

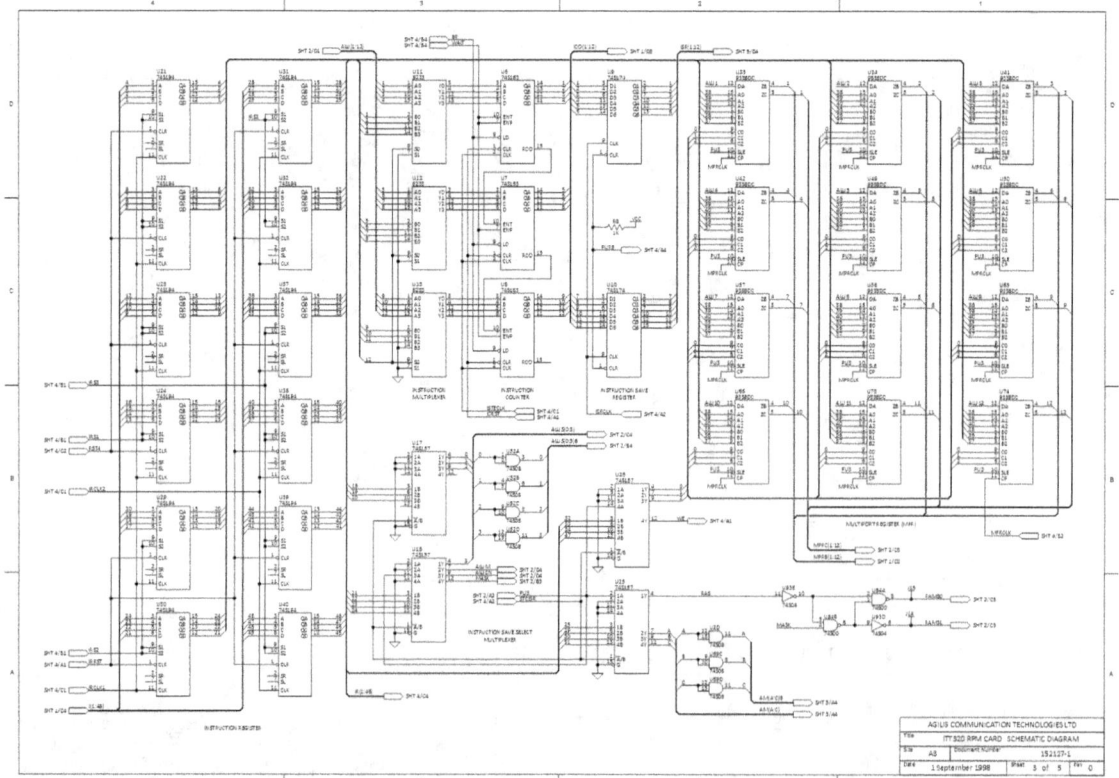

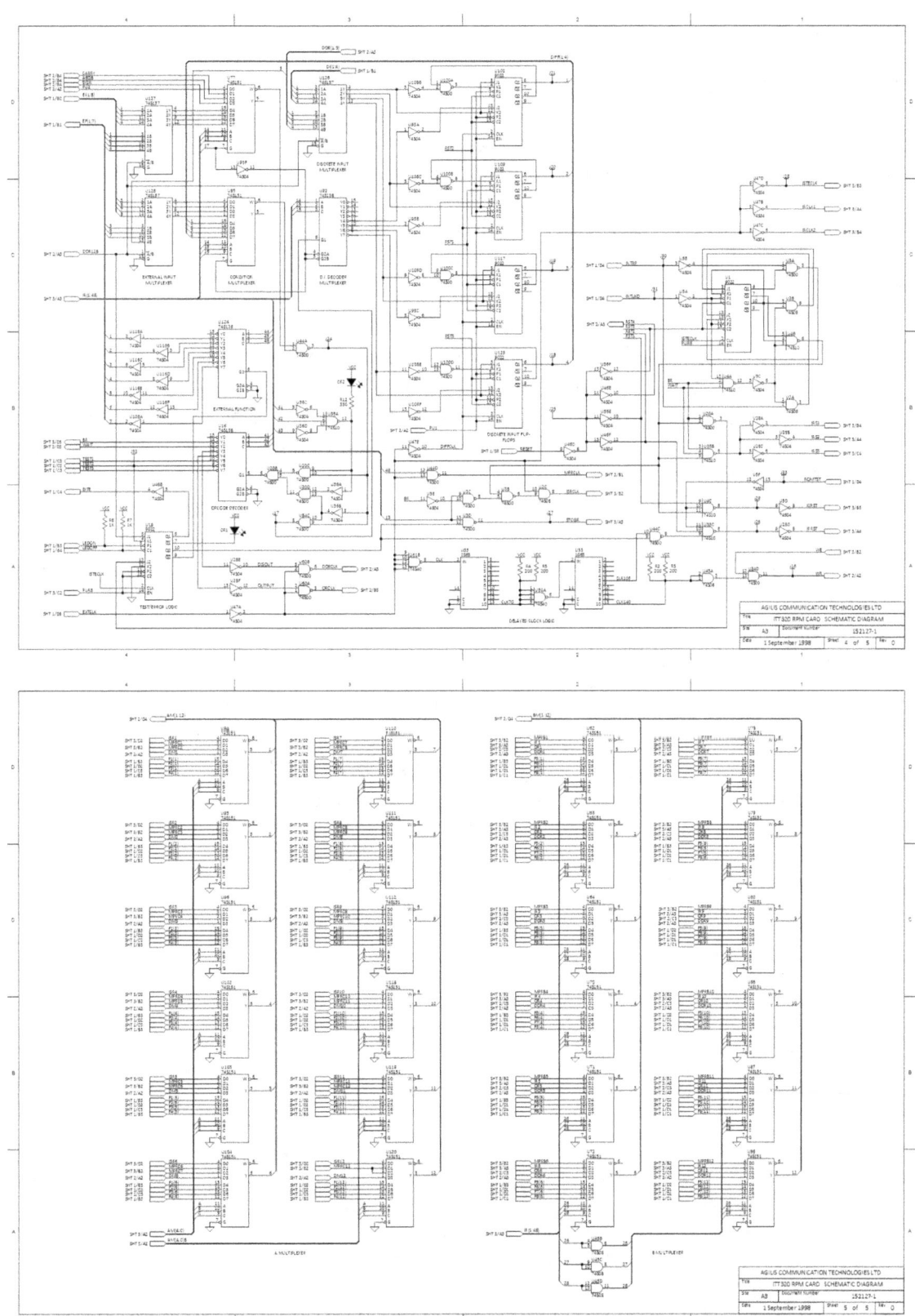

Fundamentals

After thoughts:

Twenty years on, I'm sure that equipment company would have a much-improved version in their second-generation product. The good news is, I still have the netlist of the board. The question is, if someone from that company happens to read this now, will there be any interest to give it a go with the new schematic generator?

The Defining Moment

Of course, there are numerous other topics related to PCB-RE, such as flying probes, JTAG, clip-n-learn, chip forensics and even X-ray imaging, but it is not the intention of this edition to treat each of these topics in-depth. Rather, through discussing some of the aspects relating to PCB-RE, and creating awareness in the readers of such tools and practices in existence, it is hoped that those with interest or want to pursue that level of knowledge and learning will have some idea where to begin.

In addressing the challenges, obstacles and workarounds faced by expert or apprentice alike, the inspirational words of NBA legend Michael Jordan are most appropriate here:

> Obstacles don't have to stop you. If you run into a wall, don't turn around and give up. Figure out how to climb it, go through it, or work around it.

The defining moment in any person's life is when he encounters the first challenge or obstacle that threatens to stump or stop him from a dream or pursuit. How he reacts to it will have great repercussion to his personal growth and future skills development.

Don't take my words for it; just heed MJ's advice.

TOOLS & TECHNIQUES

ScanCAD:
The Art of Perfect PCB-RE

> The art of perfection is found in the excellence of craftsmanship and a spirit of perseverance.
>
> The Author

As more companies face DMSMS and obsolescence issues,[38] there is an ever-increasing need to maintain spares and stock up for legacy systems that are operating well past their intended lifecycle. This is especially true in the large transportation, medical, automotive, aerospace and defence industries. Without a doubt, these same challenges impact other industries such as the agriculture (irrigation systems, livestock monitoring, facilities) and entertainment (musical instruments, amplifiers, theatre and stage). Often, the original manufacturer is no longer in business or no longer has the PCB fabrication, assembly and test data. In such cases, there is the urgent need to precisely regenerate needed manufacturing data from existing remaining parts, films, drawings, etc. The need for PCB-RE is very real and increasing exponentially.

When it comes to DMSMS and obsolescence with PCBs, there is the clear need for exact 'form, fit and function', so newly fabricated re-engineered PCBs will integrate properly with existing systems to avoid costly environmental and functional testing. Replacement parts that are not 100% identical to the original parts must be treated as a new design, which can be a very expensive and time-consuming proposition. The shortest path to providing operational and reliable replacement PCBs is through a high quality RE process.

There are many techniques to re-engineer PCBs and each has its distinct advantages and disadvantages. Some of the techniques covered in this chapter, and other parts of this book, are:

- Manual hand probing for bill of materials (BOM) and netlist generation,
- Optical, CT-scan and X-ray imaging systems for capturing connectivity and PCB geometry information.

---

[38] Diminishing material supply (DMSMS) is defined as the loss or impending loss of manufacturers or suppliers of items or raw materials. Obsolescence refers to a lack of availability due to statutory and process changes, as well as new designs. DMSMS and obsolescence adversely affect Product Life-Cycle Management (PLM), the overall process of managing the entire life cycle of a product from its inception to retirement.

Tools & Techniques

- Flying probe test (FPT) and bed-of-nails (BON) test systems for obtaining and validating connectivity information.
- Complimentary techniques that modify or convert data from the above processes into usable formats, and even permit information to be imported or uploaded into computer-aided design (CAD) systems, etc.

Optical, CT-scan and X-ray images of external and internal PCB layers will be reviewed along with a discussion about the pros and cons of each image acquisition process. Also, the destructive and non-destructive aspects of each of these techniques will be discussed.

The required PCB manufacturing data formats such as Gerber/drill data, ODB++, IPC-2581, etc. can be generated using some of the PCB re-engineering techniques that are presented here. Other data formats required for board testing and repair, such as netlist (IPC-D-356A) and schematics, will be covered in detail. In some cases, replacement components may no longer be available and some redesign may be needed which requires moving the data back 'up' into a CAD/CAE system. In addition, some organizations use these PCB-RE processes to 'miniaturize' existing PCBs, while maintaining existing functionality, to take advantage of new components or to meet new physical form factor or interference characteristics.

This chapter, therefore, seeks to provide a basic understanding of the various techniques for PCB-RE in support of addressing DMSMS and obsolescence as they relate to product life-cycle management (PLM).

A Legacy Crisis

The following real-life story will clearly define why PCB-RE is very relevant to our world today:

Recently, a large electrical power station supplying power to a city of over 25 million people tripped off line. The maintenance staff quickly isolated the problem and determined the cause was the failure of a small insignificant PCB (one of hundreds) in one of their control systems. They requested for a replacement PCB from their local stores inventory at the plant but the stock had run out. They immediately contacted their company-wide stores facility. Again, no stock. The emergency fall back strategy of contacting external suppliers and the original manufacturer was initiated. Bad news—the OEM was no longer in business. The system has become a legacy with no support!

What options were they left with? They must replace or repair the PCB quickly. Imagine the actual cost incurred each passing minute that this plant was down. What did they do? This organization solved their problem quickly by using the destructive, automatic, optical PCB-RE technique discussed in the following pages. Similar scenarios are playing out every day in many industries, small and large, around the world, once again clearly demonstrating why PCB-RE is not only relevant, but mandatory.

It is without question that there is an ever-increasing need for repairing and replacing PCBs in legacy systems—systems that are operating beyond their original design life, a reality that is combined with many factors that impact the ability to source spare parts. Every time the world experiences an economic roller-coaster, you can expect more manufacturers and electronic suppliers filing for bankruptcy, being acquired or fold-out. Often, critical manufacturing and repair information for PCBs are forever lost during these transitions. Data loss also occurs due to computer viruses, poor management—accidental or intentional, incomplete disaster recovery safeguards, or simply technological evolution and obsolescence. For instance, a fire broke out in one PCB fabrication facility several years ago that destroyed ALL the film archive of PCBs belonging to 150+ companies that trusted and stored their data with this company. 100% loss. No back up. Complete disaster!

The fact is, the need for recreating, re-engineering or reverse engineering manufacturing and repair data from existing parts is rapidly growing. Each industry now has third-party companies that are dedicated to repairing and supporting obsolete legacy systems such as:

- Maintaining old switch equipment for telecommunications or radio equipment for first responders.
- Test equipment for the semiconductor industry that is no longer supported.
- Transportation systems (air traffic control, train and subway switch equipment, etc.)
- Aging nuclear, hydro, geo-thermal and fossil fuel electric power plants.
- Avionics going back over a half century.
- Medical equipment that is being exported to developing countries without spares (X-ray, CT scan, ultrasound and MRI Imaging systems, etc.)
- Marine and automotive electronics.
- Military and defence systems of all sorts, etc.

Legacy[39] systems take many different forms today. Even the term 'legacy' can sometimes be surprising when referring to today's systems. Some systems may be considered out-dated or back-level just a few short years after release, leaving the end user in a difficult predicament. Massive investments in legacy equipment must be supported with spare parts that may no longer be available.

What options are there for organizations in this situation?

Ideally, legacy systems are retired and replaced prior to running out of spares. In some cases, it is possible to replace or upgrade a portion of a legacy system to 'buy time'. But more often than not, such equipment must be repaired and maintained exactly as they are.

---

[39] The Merriam-Webster Dictionary definition for 'legacy', when used as an adjective is, 'of, relating to, or being a previous or outdated computer system.' And, surprisingly, it's first known use was in 1988!

Tools & Techniques

Most complex systems are made up of a variety of inter-related and interconnected PCBs—a collection of PCBs that must 'handshake' perfectly together for a system to operate properly. Over time, one or more of these PCBs fails and needs to be repaired or replaced. In many cases, a single component out of hundreds or thousands of components on a single PCB may fail, causing the overall system or process to stop. In other instances, the substrate holding the components may have given way. Components and substrates can be damaged for many reasons including temperature, humidity, vibration, power fluctuation, insects and rodents, mistreatment, aging, accidents, repair, fire, etc. Regardless of the reason, a PCB has failed and it is now time to repair, replace, or as a final resort, reverse engineer.

PCB-RE Considerations

If a new replacement PCB (re-PCB) is desired, it is important to create an exact replica of the existing part. This is termed 'precise form, fit and function'. All original performance characteristics must be precisely duplicated. The re-PCB must have the identical electrical characteristics: crosstalk, RFI, EMI, SI, delay, etc. This is essential for the PCB to handshake properly with other PCBs in the system. In fact, the re-PCB may even need to include original manufacturing 'defects' such as drilled out traces (even traces deep down in the inner layers), jumper wires or other physical oddities that might impact the electrical characteristics. The urge to 'improve' the board design must be resisted, as any changes to the PCB, and therefore these characteristics, may create operational problems in the overall system.

In the long run, it is best to exactly duplicate the original PCB for many reasons, including:

1. Lower total PCB-RE cost
2. Time savings to bring replacement parts online
3. No recertification required (UL, CE, FCC, FAA, etc.)
4. No environment testing is needed
5. No extensive system testing is necessary
6. No expensive and time consuming complete system redesign involved
7. Improve success rate and reduce risk of failure

These benefits are derived from the fact that a PCB can be recreated that is 'identical' in all ways to the original PCB, provided the same components can be identified and sourced. This is true for single, double-sided and multi-layer PCBs. It is critical that all other elements of the PCB remain identical as well including substrate materials, dielectric thickness, conductor thickness, hole diameters, plated through hole wall thickness, solder mask material, surface treatments, coatings, etc.

In some cases, organizations may wish to improve on existing PCBs or to miniaturize or modify the PCB to accommodate new functions or handle problem areas. In other cases, the design must be modified to utilize new component packages, for example changing from a through-hole device to surface mount package, or allowing for a new style connector, new tooling, mounting holes, new form factor, correct a design flaw, improved shielding, etc. Sometimes, it is even possible to eliminate jumper wires and other patches that were added to PCBs over the years without changing the electrical characteristics of the PCB. Each endeavour to make changes to a proven PCB design is different and must be carefully managed by knowledgeable engineers.

Even if a PCB is to be improved or modified, it is always best to reverse engineer the PCB all the way back to its original design; that is, perfect form, fit and function. This will provide a strong and predicable foundation that was working and had been tested from which to base the needed modifications.

If an existing PCB is to be repaired, it is still advisable to perform a 100% reverse engineering process on the PCB. This process provides all the data needed to repair or create a 100% identical re-PCB. When this is not possible due to limited time or budget constraints, then techniques discussed in this chapter will permit a bill of material (BOM) and schematic to be generated, which are the minimum requirements needed for PCB repair.

In most cases, an RE-PCB cannot be fabricated from only BOM and schematic information, but existing PCBs can be repaired because the failing component can now be identified and replaced. Also, the repaired board can then be tested. Repairing existing legacy PCBs is a viable option when a large volume of damaged PCBs is in stock and available for repair. This is very common with aviation electronics in which parts are scavenged from airplane bone yards and used in operating aircraft. Over time, there are fewer and fewer operational spare parts while the pile of damaged junk increases.

Why not just redesign a PCB for a legacy system from scratch?

In many cases, the newly designed PCB will not play well with the other components or PCBs in complex systems. Again, the PCB must be exact form, fit and function. It must be an exact replacement and behave 100% exactly like the original PCB.

The processes discussed in this chapter do just that. Eliminating the need for recertification, system testing, environmental testing, etc. The goal is to be able to install the replacement PCB into the aircraft, automobile, medical device, electrical power station, digital piano, etc. and know with certainty that it will work as before—perfectly.

Is this being done today?

Absolutely. If the reader has flown in the last 20 years, it might surprise you to know that the global air traffic control system is, for the most part, 'legacy' equipment that is currently being maintained by the techniques which will be discussed in this chapter! Likewise, much of the legacy telephone switch equipment around the world in use today is kept running with these techniques, the newest of which include X-ray, CT-Scan and other non-destructive means and methods.

Tools & Techniques

Legacy PCB Data

When PCB-RE becomes inevitable, the types of legacy PCB data available will play a significant role in the success of re-engineering the affected PCBs. These legacy data, in many instances, may be outdated, incomplete, in bad shape or condition. Whatever the situation, having some data to work with is a helpful start; in fact, **success can be obtained even with just a partially damaged board and nothing else.**

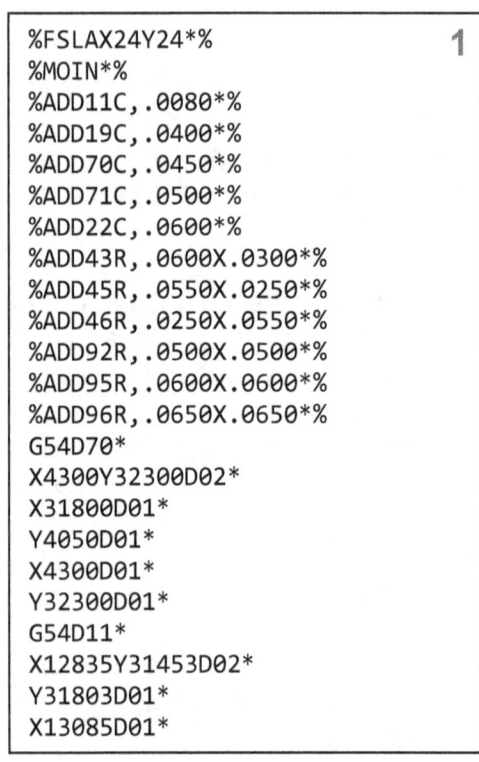
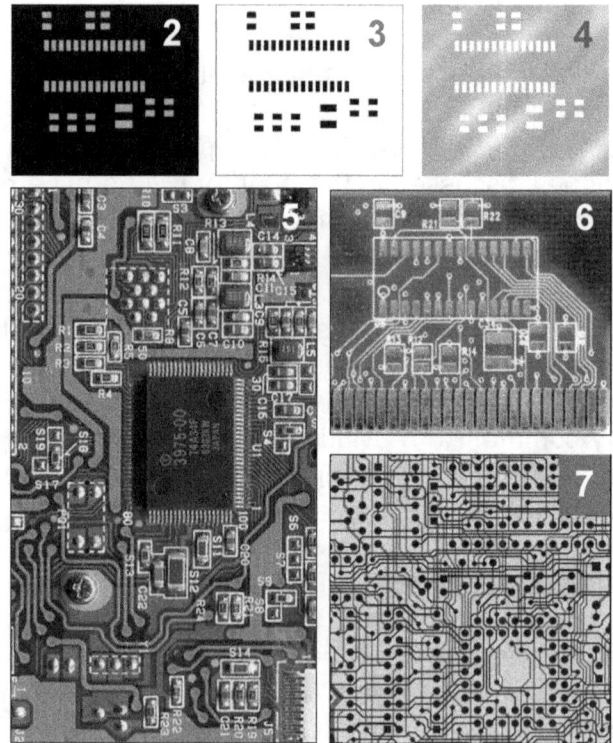

The kind of legacy PCB data being used to re-engineer PCBs today include:

1. Gerber data in 274X text format
2. Gerber images
3. Photo tools, paper drawings, microfiche, etc.
4. Solder paste, glue stencils or screens
5. Physical PCB (populated)
6. Physical PCB (bareboard)
7. X-ray or CT-scan[40] images

---

[40] CT-scan technology has only been employed in PCB-RE recently. The same 3D image data capture and post processing layering technology that has been used successfully in the medical field has now been successfully applied to PCBs, the goal being to create high quality images for each copper circuit layer without destroying the multi-layered PCB. However, this technique is still relatively new and can be subject to 'blind' spots related to lead solder and components on the PCB.

PCB data has been successfully recovered from paper drawings and, in some cases, data has even been recovered from archived low-resolution microfiche. Even non-functioning PCBs can be very useful for PCB-RE. If more than one PCB is available, additional post re-engineering validation and verification processes can be performed to confirm the accuracy of the new data. Like any RE process, it is good to have more than one source object. Indeed, the chances of recovery increases with a bigger pool of legacy data to work with.

We shall now look at the various re-engineering techniques, consider each method's strengths and weaknesses, and see how ScanCAD can integrate with some of these techniques to maximize the possibility of success.

Manual Probing

The first technique discussed is the manual probing of electrical connections on both sides of a PCB using basic test equipment such as a digital multimeter (DMM).

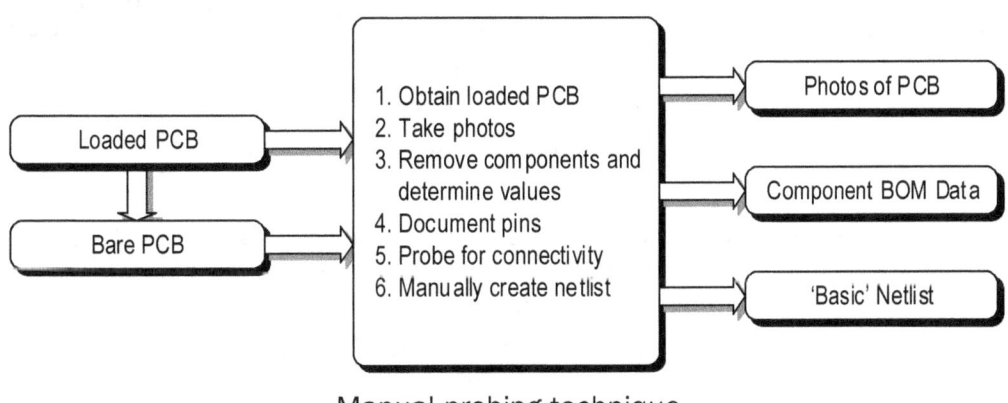

Manual probing technique

Manual probing can be easy to handle for simple boards but error-prone and time-consuming for high-density PCBs. A complex board could take many man-months of work. Validation using the resulting netlist on an electrical FPT or bed-of-nails tester is highly recommended, if not mandatory, for complex PCBs. The resulting data of this process is strictly a BOM and a netlist which cannot be used to create fabrication data for a PCB that requires identical form, fit and function. However, the data can be used to create a schematic, and therefore is very effective as a PCB repair tool.

Pros : 1. Low cost
        2. Basic netlist and BOM to create schematic

Cons: 1. Slow and laborious
        2. Error-prone for high-density boards
        3. No form, fit and function
        4. No PCB fabrication data

Tools & Techniques

Optical Imaging

This is a straightforward, destructive process which can handle PCBs of all layer counts, from 1 to 30+ layers. Output data from this process can be used to manufacture PCBs to exact form, fit and function, or it can be converted into netlist data and then exported to schematic formats, as required.

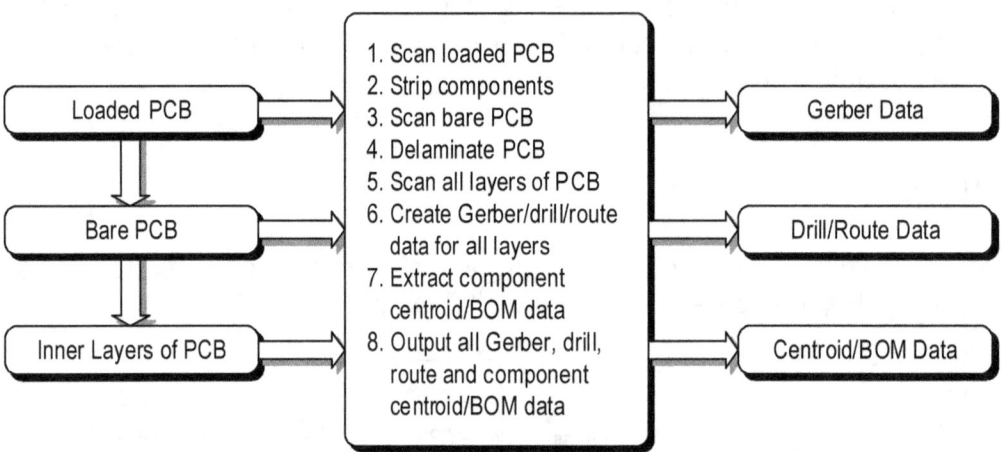

Optical imaging technique (ScanFAB and ScanPLACE)

The above flow diagram depicts the automated optical process from a populated PCB to high-quality form, fit and function PCB fabrication data.[41] All features of the PCB are reproduced to the exact same physical characteristics. This process creates exact Gerber, drill and route data files for multi-layered PCBs, supporting blind/buried vias, differential pairs, RF designs, microwave, and precise analogue circuit shape characteristics, etc. The optical imaging process ably recreates the data used to fabricate the PCB, literally making an exact duplicate right down to and including any original manufacturing defects.

With this technique, it is possible to take a visual tour down through a PCB using the images taken from each layer as the PCB is delaminated, one layer at a time, with perfect layer to layer alignment. The overleaf images show the process in action. These images were taken from a portion of a 6-layer laptop DDR memory module. The final image had bright Gerber/CAD data that were automatically generated for the inner signal layer from the color image. Note that the 'differential pairs'[42] needed for the board's high-speed timing have been accurately reproduced (see Figures 7 and 8), thereby maintaining the boards original design criteria to preserve signal integrity. Also, this technique uses high-resolution color imaging for better contrast to look deep into a PCB.

---

[41] The ScanFAB package comes with an optical imaging system—a high resolution, dimensionally calibrated scanner with a NIST certified glass calibration plate.

[42] These are 75 um (3-mil) traces and spaces.

ScanCAD: The Art of Perfect PCB-RE

Steps in the optical imaging process:

1. Bare PCB with legends on silkscreen.

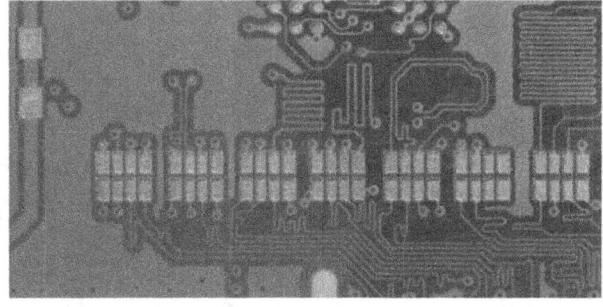
2. Bare PCB with silkscreen removed.

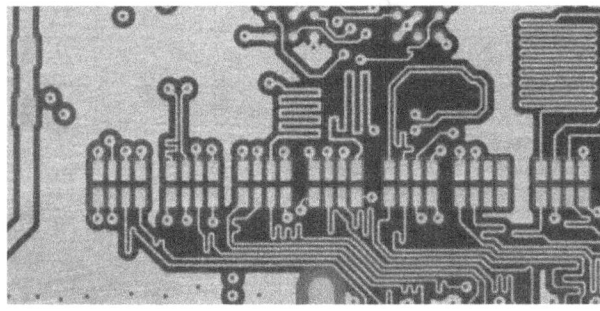
3. Solder mask removed showing top circuit.

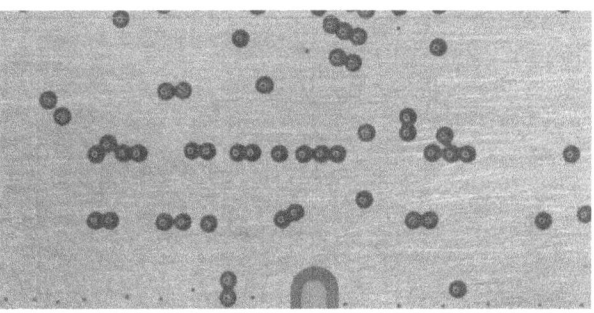
4. Top circuit removed revealing power plane.

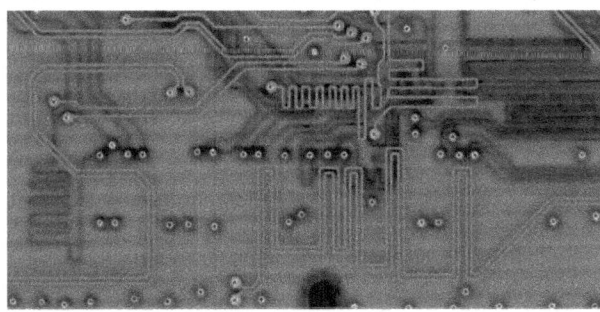
5. Power plane removed showing inner layer.

6. Next inner layer (flipped).

light-color traces against a dark PCB background

7. Inner signal layer with Gerber overlay.

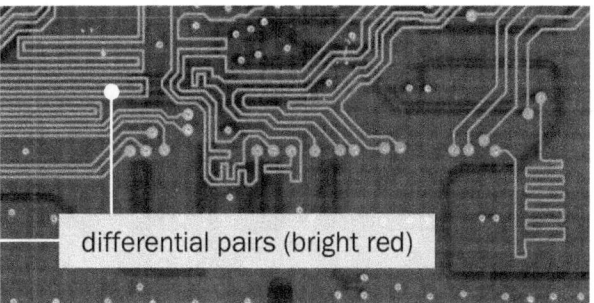
8. Flipped overlaid inner signal layer.

differential pairs (bright red)

Tools & Techniques

Figure 4 shows an image of the PCB with the top circuit layer mechanically removed revealing the second layer, in this case a power plane. Note the isolated vias. Also, note that the barrel thickness in the vias can be accurately measured. The barrel thickness as well as the exact thickness of the conductor and dielectric layers are needed when recreating the fabrication data. All of this can be captured using the optical imaging process.

Figure 5 shows an image of the power plane removed revealing the third layer of the PCB, an inner signal layer, and another vivid signal layer behind this layer. Some fiberglass dielectric materials are translucent permitting several layers to be visible all at once.

Figure 6 shows the same board flipped over, so what was a shadow in the previous image is now visible and what was visible in the previous image is now in the shadow. Note the lighter color of the conductor pattern for this layer in contrast with the darker green dielectric.

Figures 7 and 8 are the same two inner signal layers with overlaid CAD/Gerber data shown in bright red that was automatically generated to cover the traces. The new CAD data identically matches the bitmap images in terms of trace width, length, angles, etc.

Various options are available to delaminate a PCB. Chemical etching processes normally employ Ferric Chloride as an etchant but improper disposal of the residual chemical waste can pose undesirable environmental issues. Fully mechanical processes include sanding and polishing, either manually or using orbital sanders, automated CNC milling machines, or micro-milling machines specifically designed for this type of delamination work. ScanCAD offers one such micro-milling machine called the Precision Material Removal System (PMRS) for delamination and conformal coating removal.[43] It is estimated that over 95% of PCBs that are physically delaminated for precise form, fit and function globally are done using mechanical techniques.

Pros : 1. Medium cost
        2. Low skill level required to perform
        3. Fast and accurate
        4. Perfect form, fit and function

Cons: 1. PCB is destroyed

There are three PCB-RE techniques that use PCB test equipment, two of which involve the flying probe tester (bare board and populated/loaded PCB) and one using the in-circuit BON tester (bare board). These are the topics of the following three sections.

---

[43] The PMRS system was introduced recently and is primarily used by organizations processing high-density PCBs with line and space dimensions below 75 um (3-mil).

Bare Board FPT

Flying probe testers (FPT) can be used to validate the resulting Gerber/drill/netlist data created from other PCB-RE techniques (manual probing, optical, CT-scan, X-ray, etc.). For example, it is possible to use the top and bottom Gerber and drill data created from the previous optical imaging technique to run a 'self-learn' process that will generate a netlist without destroying the PCB. ScanCAD offers an FPT-based PCB-RE netlist self-learn software module that works with a variety of different bare board FPT vendors and models.

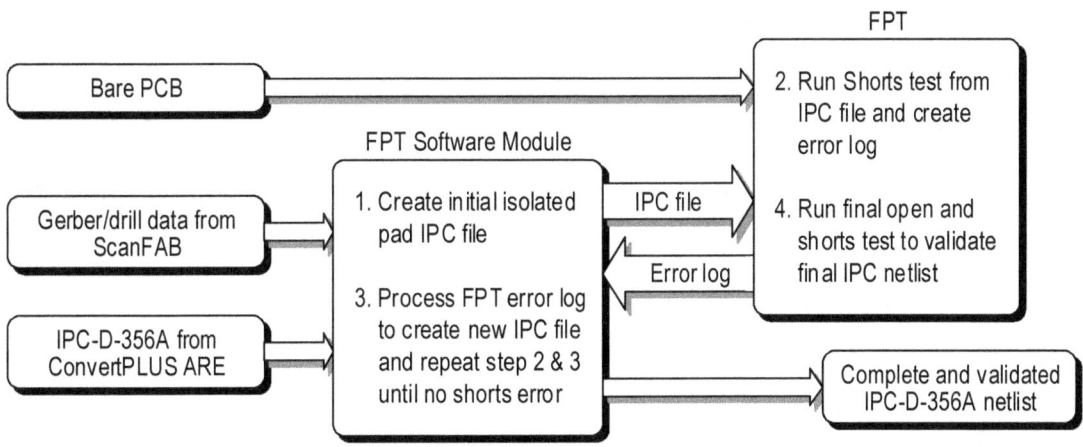

Bare board FPT self-learn technique

The obvious advantage of an FPT system is it does not require a test fixture.[44] An FPT makes use of flying probes that move in three axes to precisely make contact points on a PCB. FPT machines today can simultaneously probe both sides of a PCB, so handling a board with components mounted on the top and bottom is not an issue.

An FPT needs to have precise XY data, including pad shape, drill hole diameter and location information to operate. FPT probes are delicate and must be positioned properly on the PCB for testing to be successful. Manually locating and programming test points on a FPT machine is laborious and slow, and damage to the expensive machine or probes can happen when test points are manually located. Instead, the Gerber and drill data created from the optical imaging technique is ideal for providing the exact test locations the FPT machine needs.

Alternatively, an FPT can create a netlist from a bare board PCB using the four numbered steps shown in the above flow diagram. The FPT extracts the netlist from a PCB through a step by step process by performing a SHORTS ONLY test on the bare PCB to verify all 'isolated' pads, making contacts within the confines of the drill diameters data generated in the earlier processes. Each pad is assigned a unique label or reference ID, for example, R1 pin 1 or U23

---

[44] Conversely, an ICT bed-of-nails tester requires a custom test fixture to be fabricated, with test probes that align to the respective nodal pads on both sides of a PCB. This will be discussed in the next section.

Tools & Techniques

pin 8, etc. Capacitance test,[45] as an alternative to resistance shorts test, has also been used on FPT machines to extract the nets. It should be noted that most FPT machines use shorts testing versus capacitance testing for PCB-RE. The resulting test data is then translated into a netlist.[46]

Bare board FPT can also be used to independently validate the netlist created in earlier PCB-RE scenarios. To do this, the complete netlist generated in the earlier technique is loaded on the FPT and run as if the PCB was a new bare board being tested after fabrication, checking for both shorts and opens. It is important to note that FPT test adjacency[47] must be set to the maximum size of the PCB to ensure that all test points are checked against each other.

The bare board FPT and the next two test equipment based techniques contain a potential source of error when performing PCB-RE on PCBs. It is possible that random pads suffer from surface contamination or some physical issue that affects the electrical measurements of the FPT probes or the PCB may have experienced an internal micro-delamination.[48] For this reason, it is recommended that these electrical test techniques be supplemented by the optical imagining techniques which can detect internal opens from damaged boards to prevent this type of PCB-RE error.

Pros : 1. Non-destructive
       2. Easy to program

Cons: 1. High cost
       2. Slow and error prone
       3. No form, fit and function
       4. Netlist only

---

[45] Capacitance test assumes that each net has a unique capacitance value.

[46] Essentially the list of shorts indicates which pads are connected since they are on the same net, so the error log generated enables some form of extrapolation to determine connectivity.

[47] Adjacency is an FPT test parameter that indicates the maximum physical distance between nets represented as features on a board for a test to be run. Normally, FPT machines that are used in PCB production environments only check adjacent traces and pads, those that are close to each other since this is the only place that shorts normally take place. For PCB-RE, shorts can happen between any and all pads and traces since the PCB data is being recreated from the PCB, hence the need for a large adjacency value for the FPT process to be successful.

[48] A micro-delamination is defined as when a plated-through hole barrel may be broken when components are de-soldered and removed from a PCB, creating an open that will translate to a net not being detected by the tester. The heat from de-soldering can create gases in the inner layers of the PCB and cause an internal high-pressure gas bubble to form and break via barrels or create other opens within the PCB layers.

Bare Board BON

The in-circuit tester (ICT) is also sometimes referred to as the bed-of-nails (BON) tester because it relies on a test fixture with wired test probes that correspond to the component pads on the PCB it is testing.

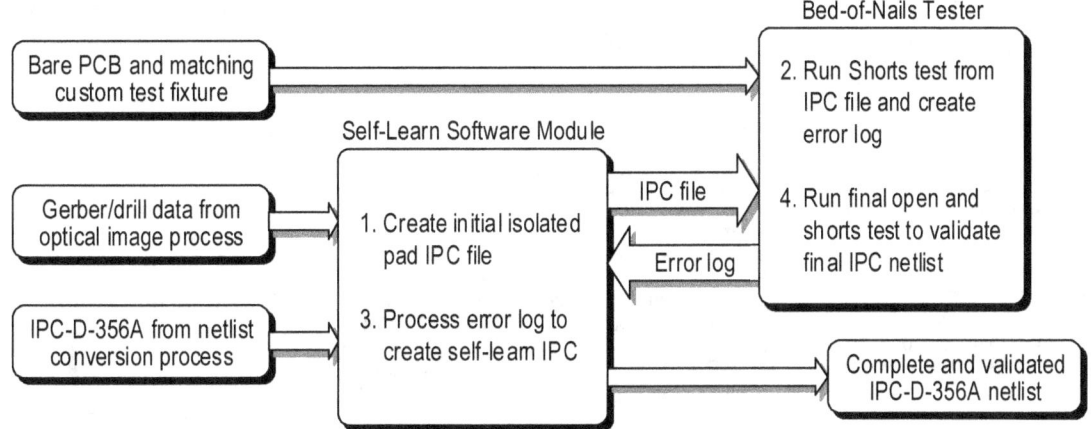

Bare board BON self-learn technique

The bare board BON technique is similar to the bare board FPT, the difference is only in the hardware platform used. And as with the FPT, it can be used to validate Gerber/drill data or generate a self-learn netlist (list of shorts). The PCB's top and bottom Gerber/drill data from the other techniques can be used to fabricate the custom test fixture required by the BON Tester. The process is non-destructive.

It is important to note that this technique requires custom test fixture to be built for each PCB. This is an additional cost in both money and time that must be considered.[49] Once the test fixture is fabricated, the validation process is much faster than FPT machines, since there is one dedicated test probe for every pad on the PCB (potentially several thousand probes) versus the limited number of probes on a FPT machine (typically 2 to 8 probes).

Pros : 1. Non-destructive
        2. Easy to program

Cons: 1. High cost
        2. Slow (if test fixture cannot be built quickly) and error prone
        3. No form, fit and function
        4. Netlist only

---

[49] A typical ICT test fixture comprising 1000 test probes on a single side (usually the solder side of the PCB under test) costs about 5-7K (USD) and takes 2-3 weeks to build, depending on the type of fixture material used. For a top and bottom fixture with 4000 test probes or more, the price can more than double.

Tools & Techniques

Loaded Board FPT

This technique follows a similar path as the bare board FPT, though the programming is much more complex given the need to test a populated PCB. In some instances, it may still be necessary to remove components when there is difficulty accessing certain test points. A thorough understanding of the populated board FPT and the PCB itself is required for this technique to be successful. These FPT machines are feature-rich and more expensive compared to their bare board counterparts, and require a higher skill level to operate as well.

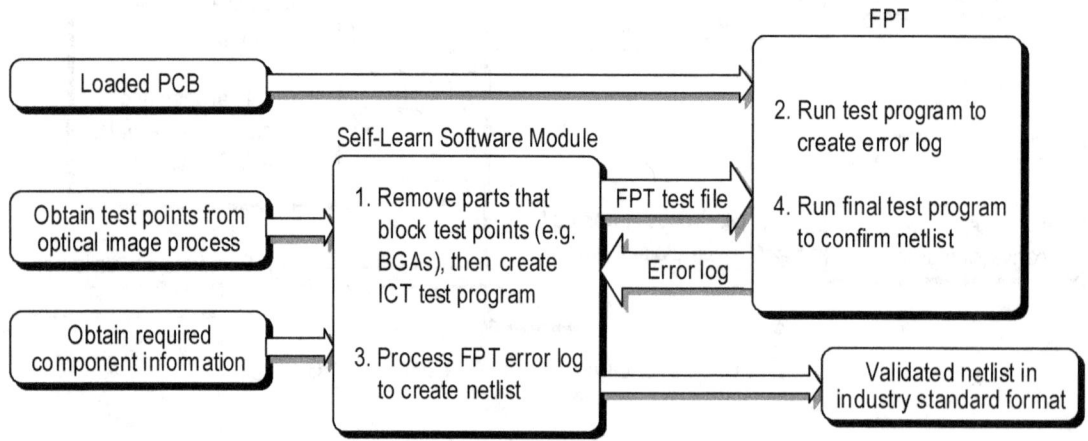

Loaded board FPT self-learn technique

There are two ways to apply this technique: first, the FPT can be used to create a netlist from a populated or a bare PCB. Using the four numbered steps shown in the flow diagram, the process can extract the netlist step by step through executing a custom created test program that understands all component characteristics on the PCB, including use of special guarding techniques to isolate certain devices or providing power to target devices, etc.

When testing populated PCBs, the probes must be able to carefully come in contact with component leads or test points that provide access to all hidden pads. Probing populated PCBs can be extremely time consuming and expensive, although possible in many cases, provided there is access to all device leads. NOTE: This can be a problem for BGA packages where all leads are hidden from the tester. A netlist can be generated from the resulting test data. Next, as with the earlier tester scenarios, the FPT can also be used to independently validate the netlist created in earlier scenarios as well as be used as a test machine for PCBs after repair.

Pros : 1. Non-destructive
        2. System can test loaded PCB

Cons: 1. Very high cost
        2. Slow and complex programming
        3. High skill level required
        4. No form, fit and function
        5. Netlist and BOM only

Optical Scanning and X-ray Imaging

This is a new technique that integrates CT-scan/X-ray imaging with the proven ScanFAB optical imaging method—the former handles the inner layers while the latter takes care of the top and bottom layers of the PCB. The ScanFAB system can import the CT-scan/X-ray images and scan the top and bottom layers of the PCB with its calibrated scanner. Using this two-prong approach, the dimensional integrity of the RE process is locked in when capturing the data from the outer (visible) layers such as legend, solder mask and non-plated hole information.

Output data from this process can potentially be used to manufacture PCBs to the exact form, fit and function, or be converted into netlist data and schematic. It could, however, be difficult or even impossible to create data for inner layers, depending on the quality and dimensional integrity of the CT-scan/X-ray images. Electrical test data validation is therefore strongly recommended.

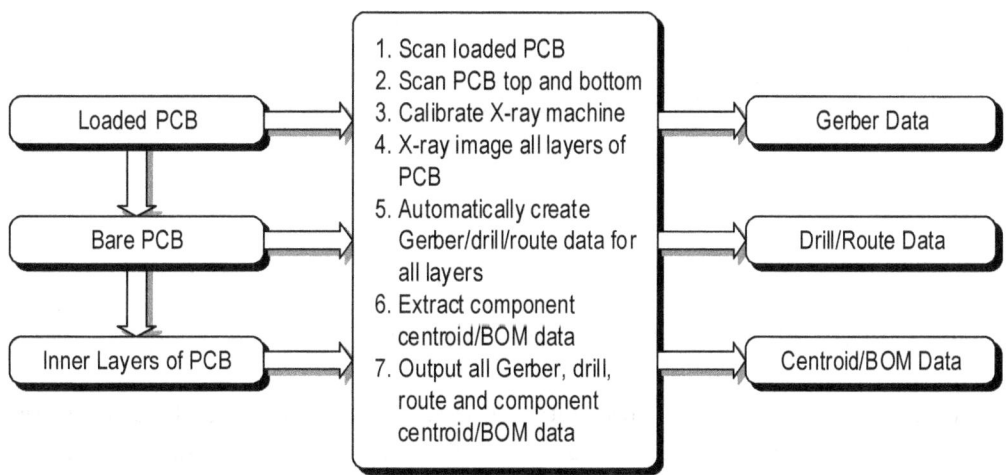

Optical and CT-scan/X-ray combination technique

There are a few issues with using X-ray systems, though:

- X-ray systems typically do not hold tight dimensional tolerances, so the inner layer images are usually 'referenced' or 'scaled' by overlaying them on the calibrated top and bottom optical PCB layer images that are dimensionally correct. The benefit of this approach is that the PCB may be reverse engineered without being destroyed to obtain form, fit and function layer design and fabrication information.

- X-ray has inherent problems regarding imaging dense materials such as lead, a problem called 'beam hardening'. Populated PCBs have components, connectors, solder and other features that absorb the X-ray energy and create 'blind' areas in the images. It may be necessary to supplement this strategy with some manual probing to confirm what is taking place in the blind areas. This is one reason why this technique

Tools & Techniques

can be error prone and requires careful post PCB-RE netlist validation using electrical test equipment.

- X-ray energy has been documented to damage or destroy certain active components. For this reason, some organizations such as NASA do not permit components and PCBs that they are sourcing to be subject to X-ray screening.

Finally, it may be difficult to obtain the layer by layer conductor, dielectric thicknesses and material types needed for precise form, fit and function using this technique, since the board is not delaminated.

Pros : 1. Non-destructive
2. Possible form, fit and function

Cons: 1. Highest cost
2. Slow and error prone
3. New technology
4. Possibility of X-ray energy damages
5. May lack physical data for PCB fabrication

Verification & Independent Validation Technique

Independent validation is a PCB-RE technique that validates the PCB data obtained using a combination of techniques mentioned earlier, depending on the number of PCBs available and whether form, fit and function (FFF) is required.

There are four possible scenarios, as follows:

1. Optical + BON* or BB FPT† with 2 PCBs (FFF required)
2. Optical + BON or BB FPT with 1 PCB   (FFF required)
3. Optical + BON or BB FPT with 1 PCB   (FFF not required)
4. Optical + BON or BB FPT + X-ray with 1 PCB

We shall now look at each scenario in turn...

*   BON – Bed-of-Nails
†   BB FPT – Bare Board FPT

1. Optical + BON or BB FPT with 2 PCBs

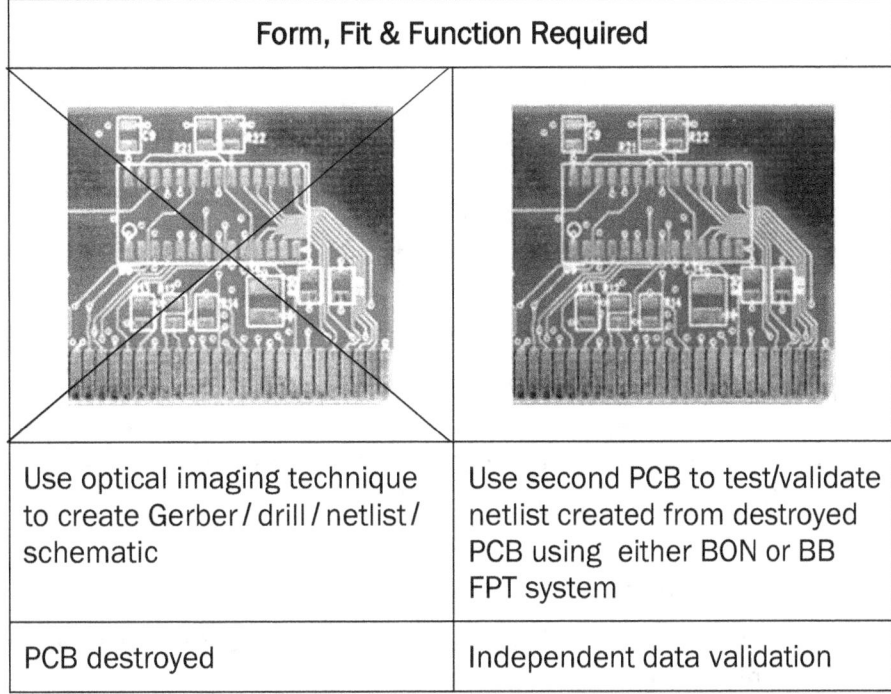

Form, Fit & Function Required	
Use optical imaging technique to create Gerber / drill / netlist / schematic	Use second PCB to test/validate netlist created from destroyed PCB using either BON or BB FPT system
PCB destroyed	Independent data validation

If it is possible to obtain two PCBs, this is the most desired combination for PCB-RE and validation. Two or more PCBs are more than likely available for PCB-RE. It is very rare that only a single PCB remains for PCB-RE. The first PCB is used to create all the needed data. The second PCB is used to independently validate the netlist that was created from the first PCB. If there are any discrepancies, this technique permits the ability to validate connections in question and identify any possible errors by reviewing the high quality optical images that have been captured for every layer of the PCB. The first PCB has been converted into a visual data bank or image source for all layers. These images are high quality, calibrated and represent ALL layers of the PCB. The digital data can be archived and used as needed in the future.

Pros : 1. Fastest method
2. Least effort
3. Independent data validation
4. Safe, accurate and preferred
5. Supports form, fit and function

Cons: 1. Minimum two PCBs required
2. One PCB is destroyed

Tools & Techniques

2. Optical + BON or BB FPT with 1 PCB

Form, Fit & Function Required

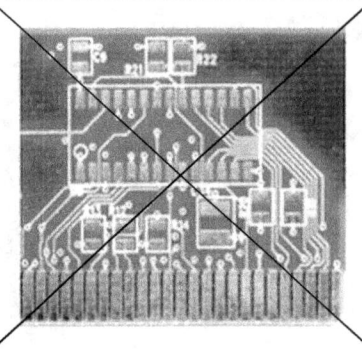

- Use optical imaging technique to program FPT process for top and bottom layers of PCB.
- Use either bare board or in-circuit FPT to generate the first netlist on the FPT system.
- Then, use optical imaging technique to create full Gerber/drill data for all layers and create second netlist.
  PCB is destroyed in the process.
- Finally, compare the two netlists.

This technique is the most time-consuming but the quality of the data is high, especially if there is only one remaining PCB. Two independently created netlists are used to validate the data for a single PCB. The first netlist is created using the FPT technique discussed earlier. Once the netlist is created and validated on the FPT machine, the PCB is then processed using the automated optical imaging technique that destroys the board to obtain images of all layers. A completely independent netlist is then generated from this optical imaging data. Both netlists are compared to confirm that they match.[50] Again, if there are any discrepancies, the solid image data for all board layers is used to resolve potential errors.

Pros : 1. Safe, reliable
        2. Life-saver when only one PCB is available
        3. Independent data validation
        4. Supports form, fit and function

Cons: 1. Slow, requires much effort
        2. PCB is destroyed

---

[50] There are a variety of software tools that can be used to compare netlists. The ConvertPLUS product from ScanCAD is one such tool.

3. Optical + BON or BB FPT with 1 PCB

Form, Fit & Function Not Required
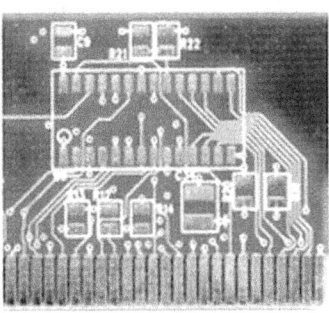
<ul><li>Use optical imaging technique to program FPT or create fixture for BON tester.</li><li>Use automatic probing and self-learn software module to generate netlist.</li><li>Use automatic probing to perform final netlist verification, not independent validation.</li></ul>

This is the recommended technique when the only goal is to obtain a netlist, presumably to support PCB repair activity. If possible, obtain more than one PCB and remove the parts from one and use the bare board PCB-RE technique. If it is not possible to remove the components, then use the populated PCB-RE technique. Be sure to validate the resulting netlist data by carefully testing the PCB using the respective FPT machine.

Pros : 1. Works with one or more PCBs
        2. PCB is not destroyed
        3. Able to extract netlist and create schematic for repair

Cons: 1. Slow and error prone
        2. High cost
        3. No form, fit or function data
        4. Verification only, no independent data validation

Tools & Techniques

4. Optical + BON or BB FPT + X-ray with 1 PCB

Form, Fit & Function Required

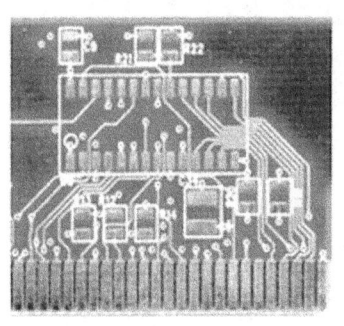 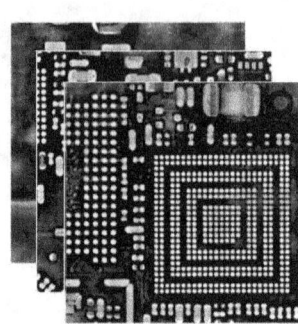

- Use optical imaging technique to obtain calibrated images of the top and bottom of PCB.
- Next, use X-ray or CT-scan tools to obtain images of all the copper layers. Then merge the optical and X-ray images to create full Gerber/drill data for all layers, add component information and create netlist. PCB not destroyed in the process.
- Finally, electrically validate the netlist on an FPT system using the same PCB.

Very few organizations can make the significant investment needed for this technique but if budgets permit, this is the recommended solution if the PCB cannot be destroyed for any reason. The hardware and software cost can reach over USD $2M for this solution. If there is one board left in the world, and the decision has been made that this board cannot be destroyed, then this is the best solution to move forward with PCB-RE. This is a combination of the optical imaging method and CT-scan X-ray based technology to obtain all PCB layer information. It is strongly recommended to consider performing an electrical test using a populated PCB on an FPT machine to validate the resulting data.

Pros : 1. Works with a single PCB
        2. PCB is not destroyed
        3. Possible form, fit and function data
        4. Independent data validation

Cons: 1. Slow and error prone
        2. Very expensive
       3. New technology
       4. Potential damage to active devices

Netlist Data Conversion

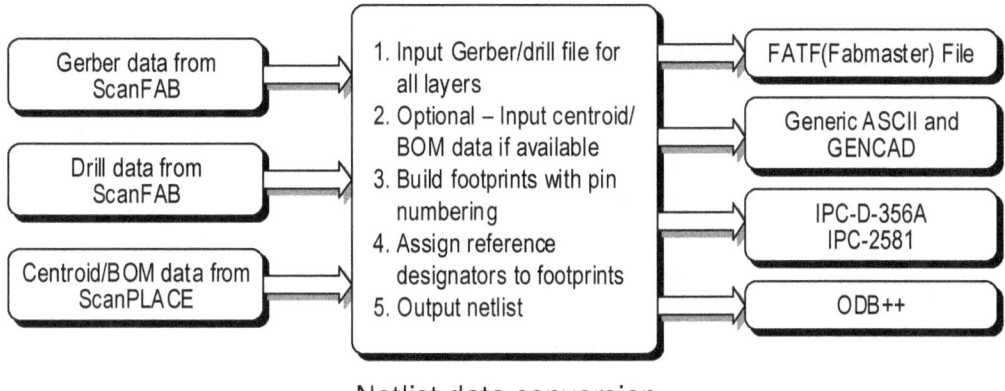

Netlist data conversion

When Gerber, drill and BOM data has been created from the various techniques discussed, it can be converted into 'rich' netlist formats to be used for PCB fabrication and assembly. Data can also be input for schematic generation and/or be used as input for the electrical testing processes mentioned, that is, offline programming for either bare board or populated board using FPT testers or creation of fixture fabrication data and test programs for ICT testers.

The flow diagram illustrates how data from earlier processes are enhanced using ScanCAD's ConvertPLUS ARE[51] data conversion capability to add component pin numbering, footprint creation and a variety of netlist outputs as shown. There are a variety of other tools on the market that provide this capability also.

IPC-D-356 is a standardized netlist format, of which 'A' describes the electrical test format. It is not a Gerber format but contains information about the test points and net connections on a board to provide a standard set of information for bare board testing.

The IPC-2581 format is a relatively new IPC industry standard for a comprehensive and rich netlist format. This was released as a vendor neutral option to the previous proprietary formats that are in common use today.

ODB++ is a proprietary CAD-to-CAM data exchange format used in design and manufacture of electronic devices. Its purpose is to exchange PCB design information between design and manufacturing and between design tools from different EDA/ECAD vendors.[52]

---

[51] Advanced Reverse Engineering

[52] Both IPC-2581 and ODB++ have the same amount of information necessary to accurately reconstruct the PCB. IPC-D-356 won't tell you anything about the trace widths, component orientations, special board features, etc. It is strictly for testing net connectivity.

Tools & Techniques

Schematic Data Conversion

Netlist and BOM data can be used to generate schematics and possibly even the complete PCB layout in a CAD/CAE environment. The following flow diagram shows how data from earlier steps are merged and manipulated to create high quality schematics. While there are a variety of tools on the market that provide this capability, ScanCAD offers two options: a simple low-cost package as well as a feature rich, more expensive package. The more expensive package permits importing component libraries from and generating schematic outputs that are compatible with existing CAD/CAE systems. All components listed in the BOM must already be found in the schematic generation package prior to processing. This is mandatory for the generation of symbols, pin numbering and graphics needed in the final schematic.

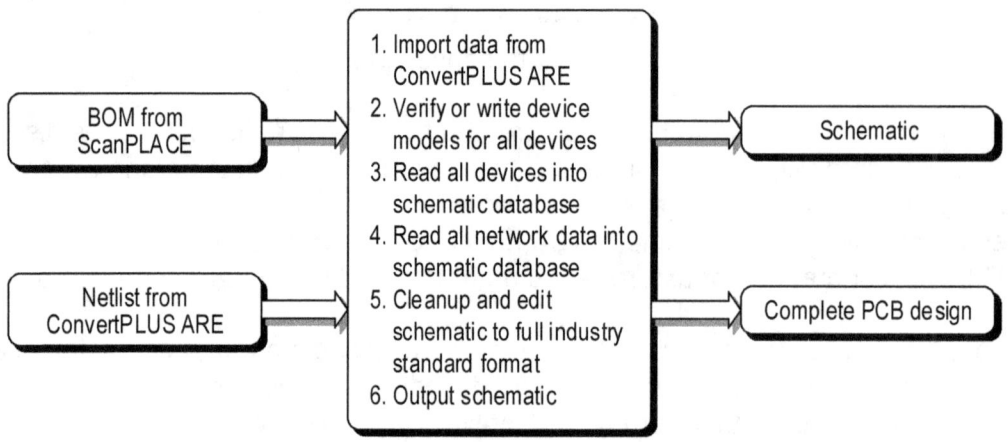

Schematic data conversion

The following schematics show what is possible using these techniques:

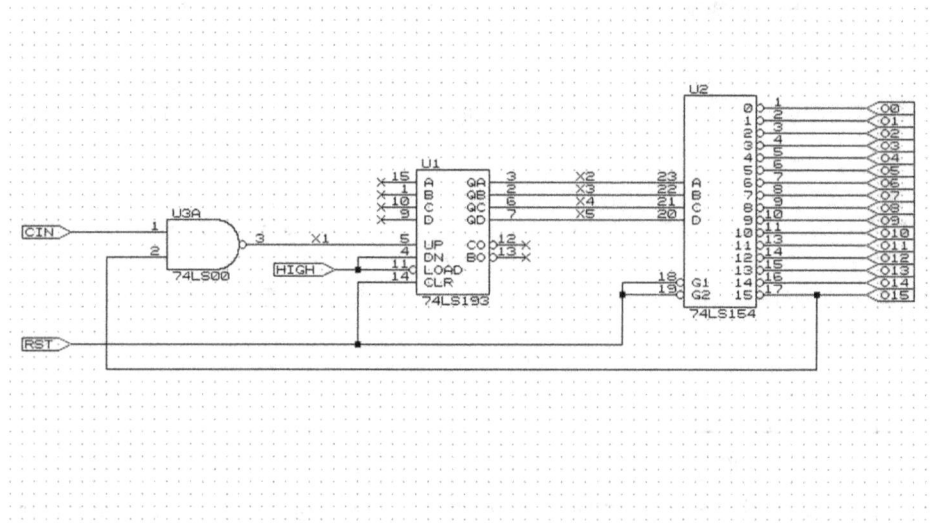

A simple schematic diagram

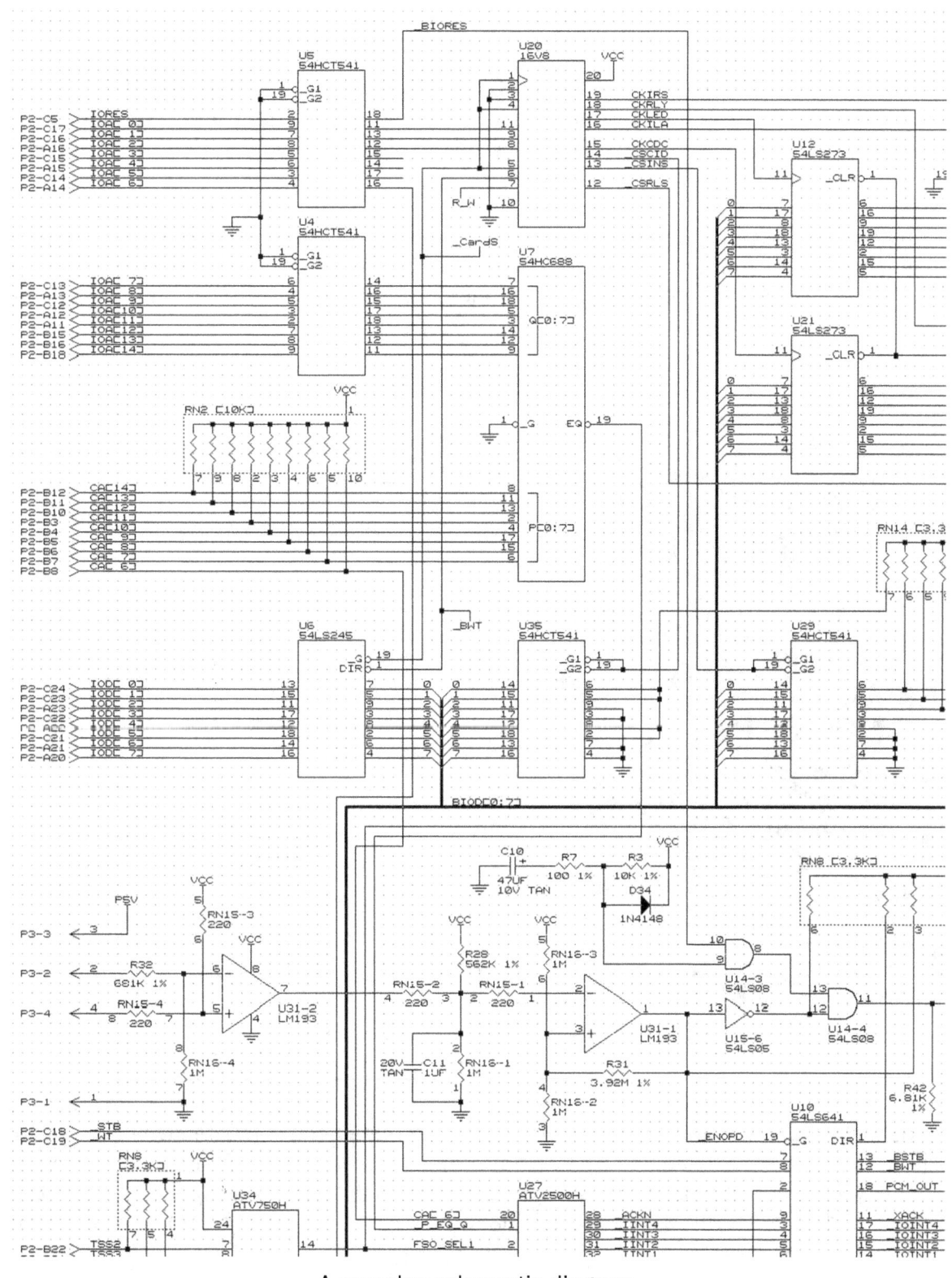

A complex schematic diagram

Tools & Techniques

Conclusion

The need to manage (repair and replace) PCBs in legacy systems is increasing globally at a rapid rate because many existing systems are being operated well past their intended lifecycle while spare parts are becoming harder to find and design and manufacturing information are no longer available.

Fortunately, there are several proven techniques that can precisely regenerate or re-engineer manufacturing data from an existing pool of legacy PCBs, photo tools and other input sources. Both destructive and non-destructive techniques are options to consider but cost will most likely be a major deciding factor.

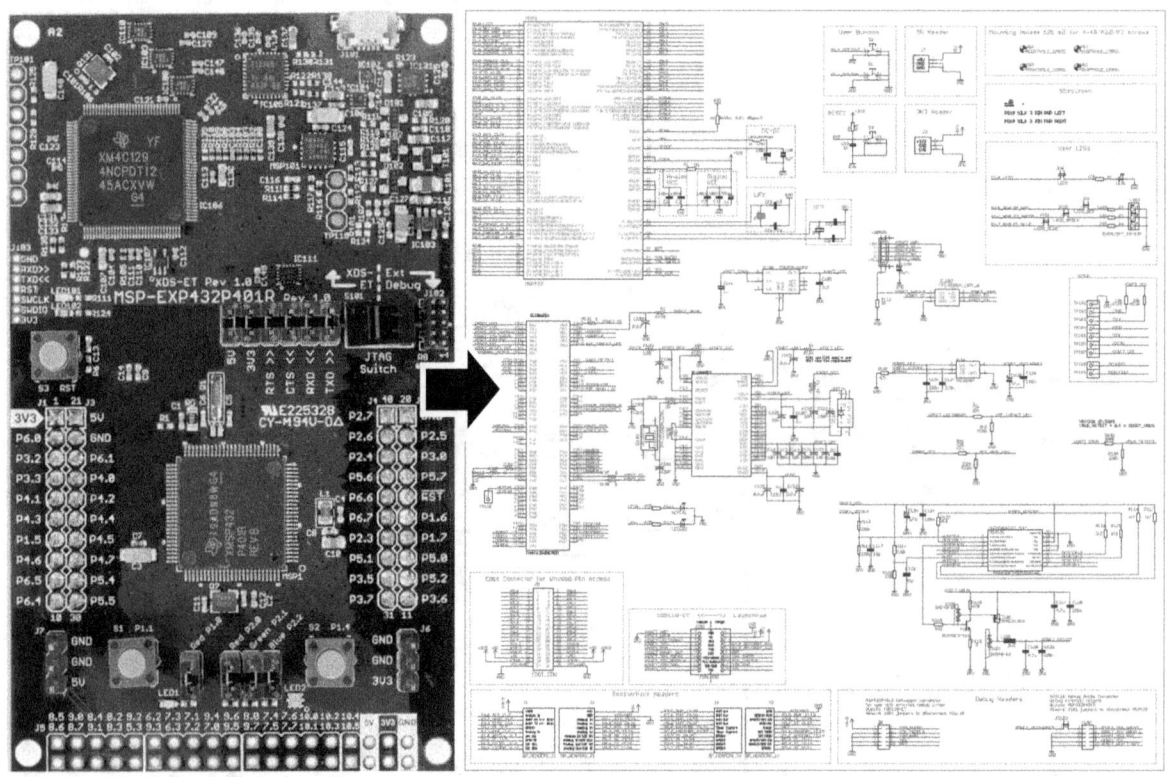

From PCB to schematic diagram.

If a complete BOM and correct netlist can be re-created, it is then a step away from generating high-quality schematics and converting to CAD data to repair existing defective PCBs, with the ability to fabricate replacement PCBs that have the exact same 'form, fit and function' needed to properly integrate into these critical legacy systems. In this way, new replacement PCBs can be fabricated as exact replicas of the original parts to resolve the problem of obsolescence and diminishing supply.

ScanCAD: The Art of Perfect PCB-RE

In summary, the following techniques were covered:

- Manual hand probing for netlist generation.
- Automated optical imaging of all PCB layers using delamination processes to create perfect form, fit and function manufacturing data to fabricate replacement parts and generate netlists.
- Flying probe test (FPT) and bed-of-nails (BON) test systems for recreating PCB netlists and validating connectivity information.
- CT-scan and X-ray imaging systems for capturing connectivity, netlist, board geometry and manufacturing information without delaminating or destroying the PCB.
- Converting the Gerber, drill and BOM data created from legacy PCB layer image data into industry standard netlist formats such as ODB++ or IPC 2581.
- Finally, creating high-quality schematics from the BOM and netlist information.

A combination of the above techniques will provide the most thorough, cost effective and productive solution, but ultimately budgets and time constraints will dictate which technique best suits an organization's needs.

We cannot stress enough the need to VALIDATE or VERIFY re-engineered data. There are various methods to perform this validation, based on available PCBs and access to electrical testers. However, INDEPENDENT validation of the data is strongly recommended. Surprisingly, this is possible even with just a single remaining PCB. Creative use of combined techniques can, in fact, allow a single PCB to produce two independent netlists for comparison as well as for data validation.

Organizations can now move forward with confidence when maintaining and supporting legacy PCBs for their mission critical systems. Coming back to the story of the power plant in the chapter introduction: They used a combination of techniques[53] to resolve their emergency issue and were up and running again in less than 72 hours. The hard fact was, they had to repair or replace the legacy PCB at all cost. The good news is there are several proven and reliable choices from which to choose from, back then for them, and more so today for anyone who needs it.

---

[53] Specifically, the automatic optical imaging technique with independent data validation.

PCB-RE: Tools & Techniques

Tools & Techniques

William (Bill) Loving, BSME, founded ScanCAD International, Inc. in 1990 using his extensive electronics manufacturing industry experience gained with IBM and robotics company, OZO Diversified Automation, Inc., and his formal engineering education. His main focus is on customer-oriented technical applied technology solutions, and heading ScanCAD's product leadership in support of today's dynamic manufacturing challenges, especially in the areas of obsolescence and diminishing supply as they relate to legacy PCB reverse engineering.

Bill has spoken globally on various subjects such as PCB reverse engineering, process-control using low-cost inspection systems, interactive paperless electronic work instructions and reducing complexity in today's manufacturing environments.

Jeff Rupert, BSEE, MS Optics, is the Director of Sales & Business Development for ScanCAD International, Inc. He has over 24 years of experience in the electronics industry and another 7 years in other technical fields. He is also responsible for the company's global sales, marketing and business development for process control solutions including PCB Re-Engineering systems, inspection systems, and interactive paperless electronic work instructions.

Jeff has authored several technical papers on a variety of subjects including 3D optical scanning products for solder paste and component placement inspection, flexible multi-purpose inspection systems, electronic work instructions and optical systems and components.

ScanCAD's portfolio of process control tools include:

- The #1 selling PCB re-engineering systems in the world since 1990
  www.PCBreverseEngineering.com
- The #1 selling stencil/emulsion screen inspection systems in the world since 1998
  www.STENCILinspection.com
- The #1 ranked Work Instruction Software in the world since 2009
  www.WorkInstructionSoftware.com

Addendum

The ScanCAD PCB Re-Engineering System comprises two components: the ScanFAB System and the Precision Material Removal System (PMRS).

The ScanFAB System includes both hardware and software components. The hardware is a high-resolution dimensionally calibrated scanner for accurate image acquisition, and comes supplied with a NIST[54] certified glass calibration plate and some scanning accessories. The ScanFAB software processes the scanned images to create the Gerber/drill data.

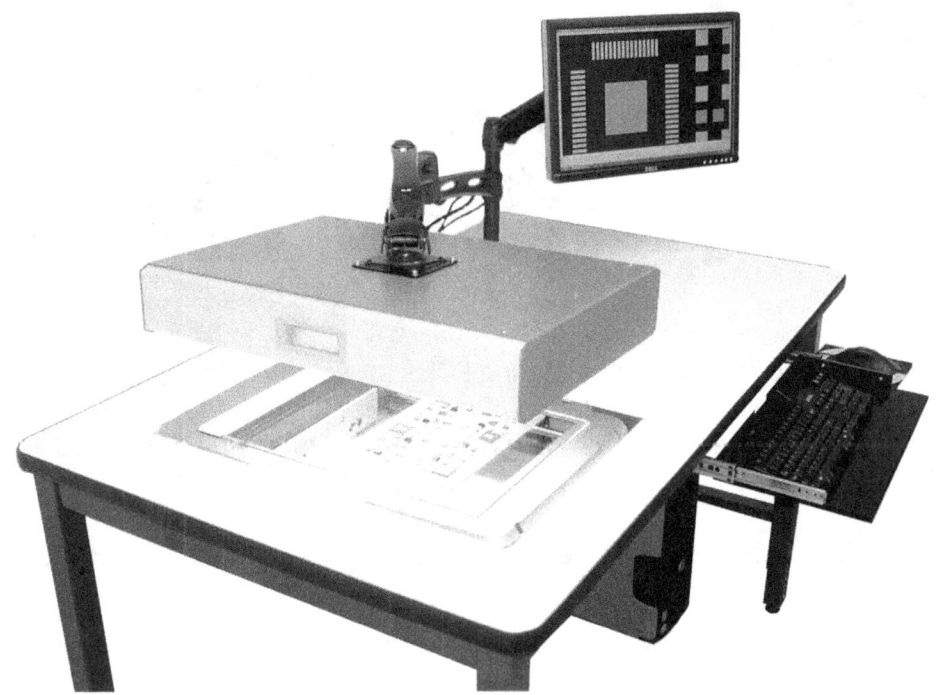

The scanner sits on a shelf below the desk surface so that the scanner surface is flush with the desk surface. Mounted to an articulating arm is a transmissive light which is very useful for imaging certain parts such as film, stencils, etc. The flexible scanner platform allows for image acquisition in either color, greyscale or B&W, at a variety of resolutions depending on the smallest feature size and the type of part being scanned. The scanner is A3 size, but there is no limit to the size of part that can be scanned—larger parts are simply processed by performing multiple scans and the software can stitch the data together into one large image.[55]

---

[54] National Institute of Standards and Technology

[55] The workstation desk shown above is an optional item. The open design of the desk lends itself to scanning parts larger than the A3 size of the scanner.

Tools & Techniques

The PMRS system provides a very effective solution to delaminate PCBs by removing material in a very precise manner, one layer at a time, with minimal risk of damage to inner layers, avoiding lost data or incorrect data generation. The system is fast, easy to use, safe and provides very high-quality results.

The system can be used for both delaminating PCBs as well as removing conformal coatings. It comes in a standard version for delamination purpose alone, and an ESD-control version for preventing damage to static sensitive devices when removing conformal coating from populated PCBs.

Basically, the PCB re-engineering process involves iterations between these two hardware platforms:

1. Scan one side of the loaded board on the ScanFAB system as the first layer and include a layer description.
2. Scan the other side of the loaded board as another layer and include a layer description.
3. Remove all components.
4. Scan top side of the bare board as another layer and include a layer description.
5. Repeat step 4 for bottom side of the bare board.
6. Remove a layer (silkscreen, solder mask or conductor) using PMRS and/or other techniques.
7. Repeat steps 5 and 6 until images of all layers, including inner layers, are captured, labeled and perfectly aligned in the ScanFAB system.

After scanning, board delamination, and layer to layer alignment are completed, the ScanFAB and ScanPLACE software is used to generate Gerber, drill, centroid, BOM, etc. data.

The following screenshots outline the process of creating the PCB data from its images:

1. Color image of top side of bare PCB scanned at 2000dpi—before Gerber has been placed using ScanFAB's automatic vectorization functions.

2. Same image showing 'monochrome' image of only the gold pads (results of ScanFAB Color Separation Process). The monochrome image is required for vectorization.

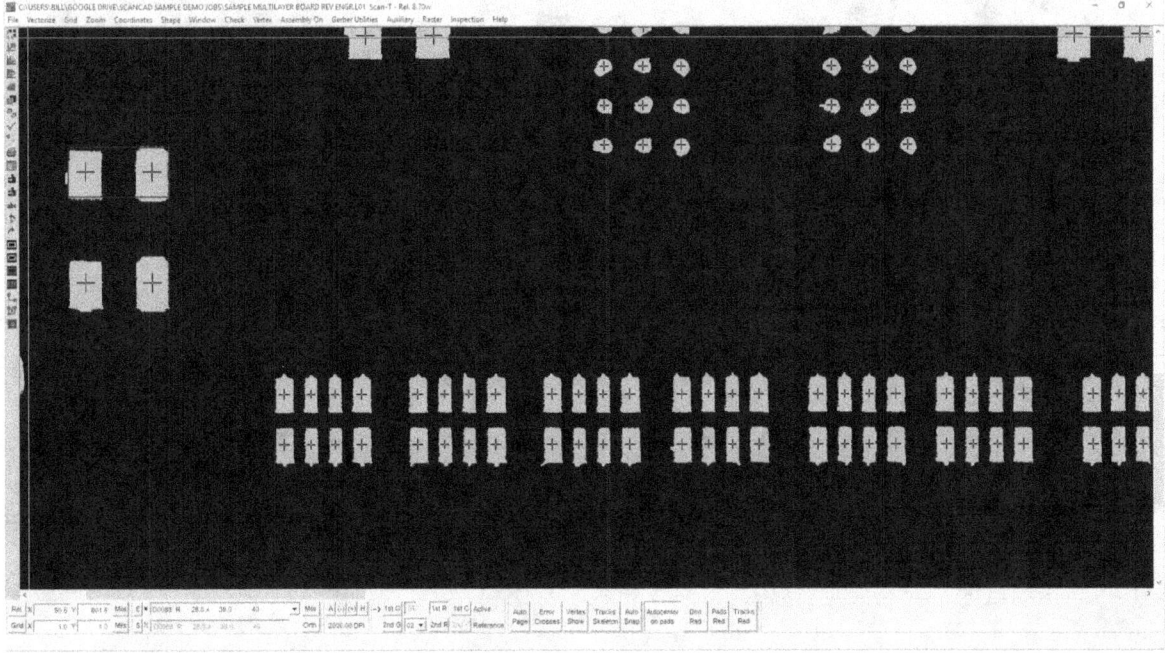

Tools & Techniques

3. Same image showing a cursor with a rectangular aperture over a monochrome raster pad image. The operator can manually edit any and all Gerber data as needed with the benefit of the dimensionally correct image information.

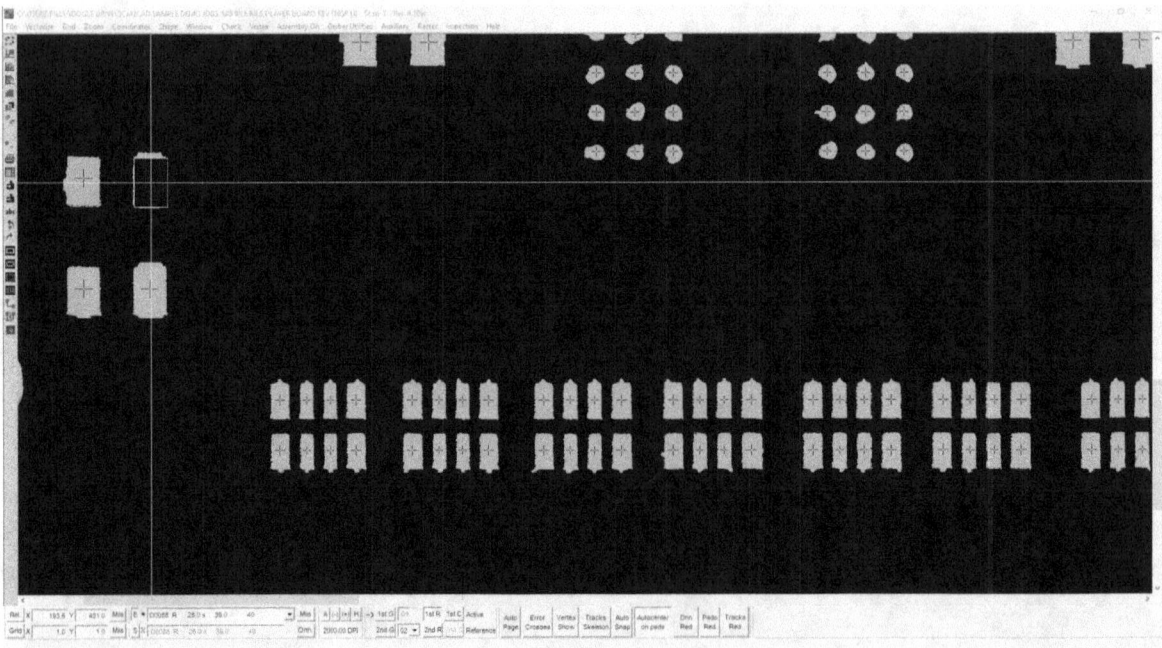

4. Same monochrome image after Gerber data has been automatically placed over pads.

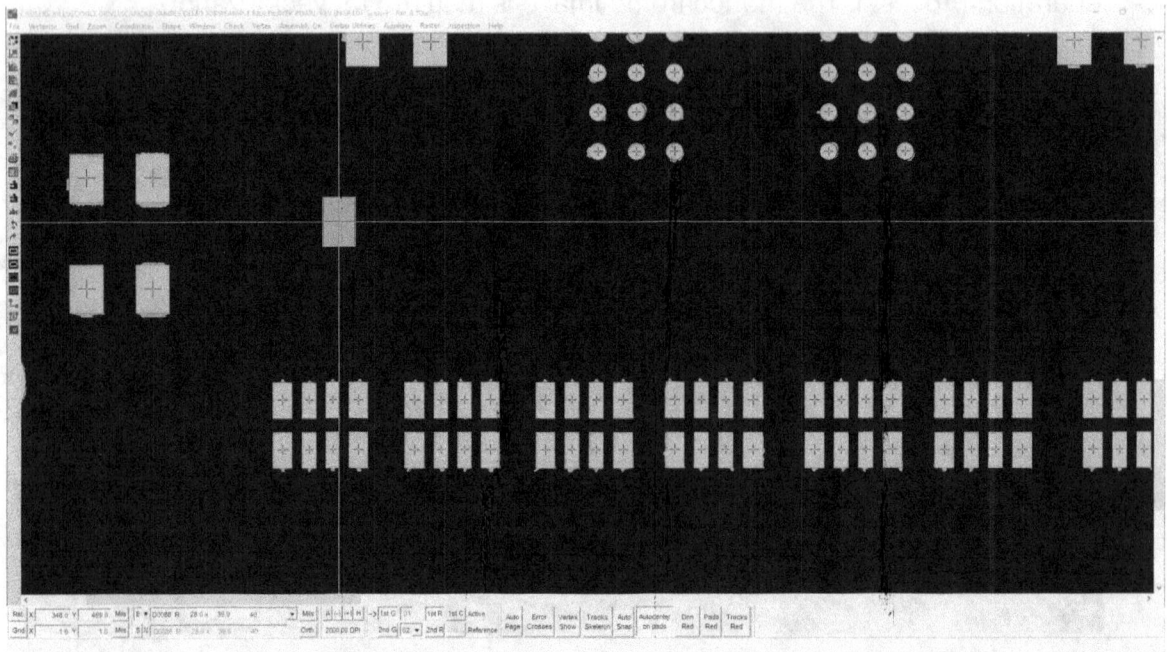

5. Same color image with GERBER data. Gerber data is ready for output for this layer from ScanFAB. Gerber can be output in 274X or D formats.

6. Same image of Gerber overlay with cursor showing a rectangular pad from the component C26.

Tools & Techniques

7. Same image with ScanPLACE functionality activated, showing an additional row of data at the bottom of the screen. Note the row with component centroid, rotation, REF ID, package ID, etc. Also, note that the cursor is snapped on C26 showing component information for this 0805 SMD in fields at the bottom of the screen.

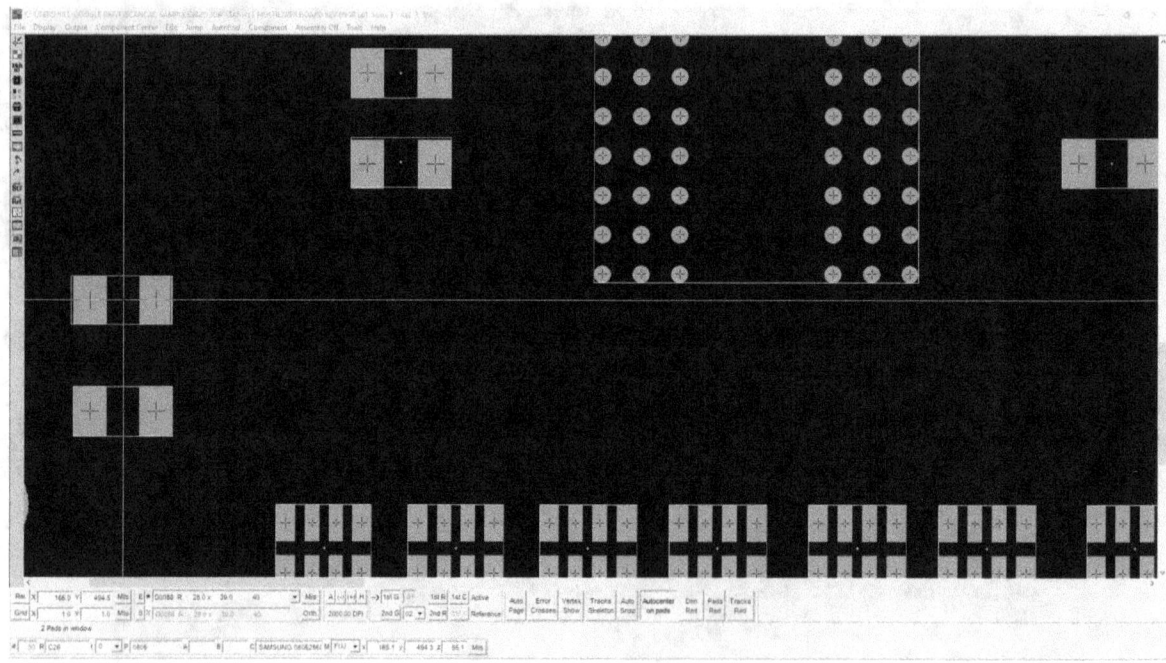

8. Note the data for C26 is also available in a table format, versus overlaid on image, for easy viewing and editing. The operator can edit the data in either environment, graphic...

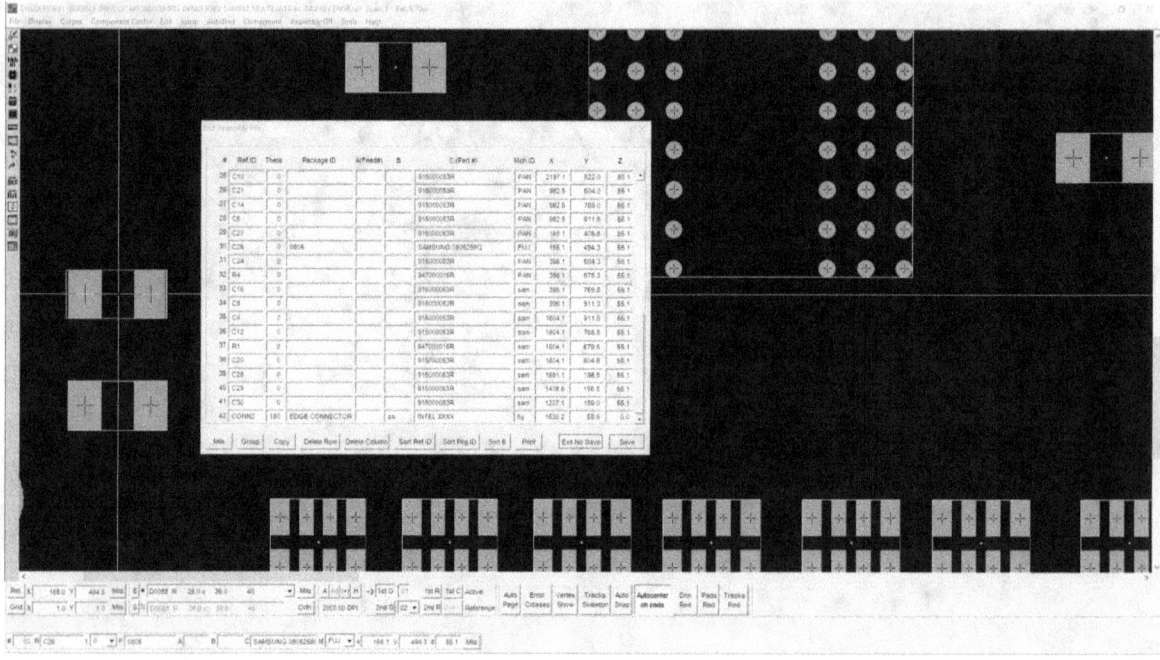

or table...

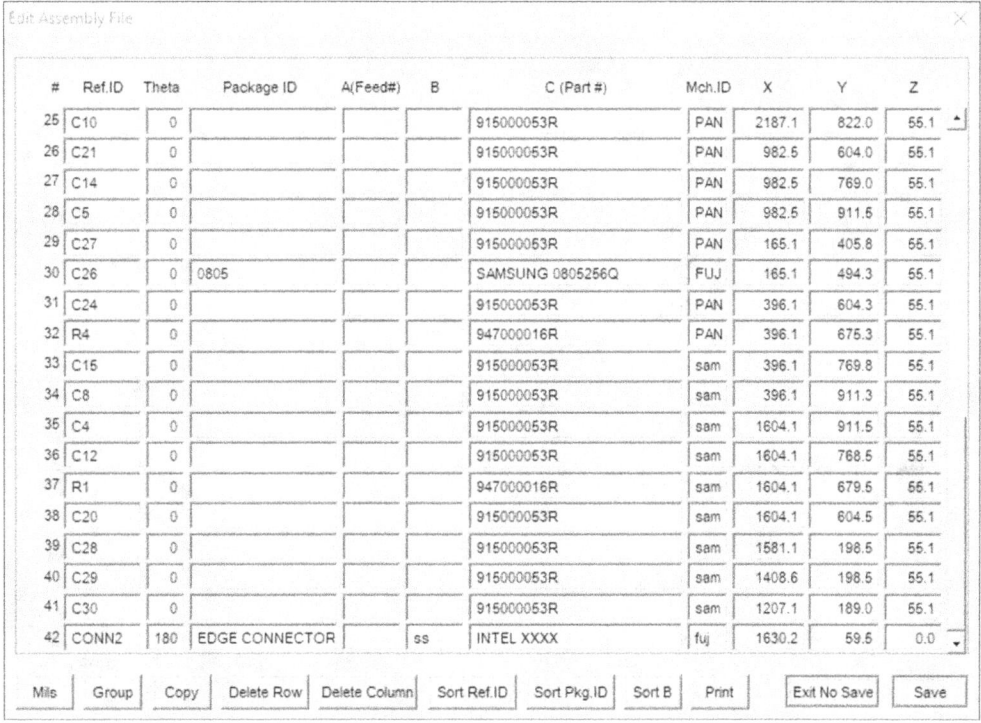

9. Same image now in color with component data showing as gray windows over color image, BOM data is ready to be output through ScanPLACE to ConvertPLUS ARF for pin number and preparation for Schematic Generation. Note C26 in the color image...

Tools & Techniques

... and in the generic output text file shown below. ScanPLACE has over 70 output formats for this PCB assembly data.

```
SAMPLE MULTILAYER BOARD REV ENGR.PPF - Notepad
File  Edit  Format  View  Help
Ref.ID. Theta   "X"    "Y"    "Z"   Mach.ID  Pkg.ID.      A(Fd#)  B    C(Part#)

F        0      135    845    0              FIDUCIAL
F        0      6574   974    0              FIDUCIAL
F        0      6485   2804   0              FIDUCIAL
U1       180    4813   1869   0     fuj                           ss   938051XXXR
U3       180    1753   1869   0     fuj                           ss   938051XXXR
U5       270    6243   1246   0     fuj                           ss   933000313R
RN1      0      5582   754    0     fuj                           ss   947000152R
RN18     0      845    754    0     fuj                           ss   947000152R
RN15     0      1674   754    0     fuj                           ss   947000152R
RN14     0      1965   754    0     fuj                           ss   947000152R
RN13     0      2236   754    0     fuj                           ss   947000152R
RN12     0      2546   754    0     fuj                           ss   947000152R
RN8      0      3626   754    0     fuj                                947000152R
C21      0      2496   1534   140   PAN                                915000053R
C14      0      2496   1953   140   PAN                                915000053R
C5       0      2496   2315   140   PAN                                915000053R
C27      0      419    1031   140   PAN                                915000053R
C26      0      419    1256   140   FUJ      0805                      SAMSUNG 080525
C24      0      1006   1535   140   PAN                                915000053R
R4       0      1006   1715   140   PAN                                947000016R
```

10. Same original color image is now processed to capture legend (silk screen) using ScanFAB color separation processes. Instead of looking for gold pads, the function is now looking and color separating the white legend information.

11. Same image with color image turned off and only the new monochrome image for the legend is showing. Each of these monochrome images are stored on different layers, at least one image for each Gerber layer of the PCB—solder mask, legend, pad master, circuit layer, ground/power layer, etc.

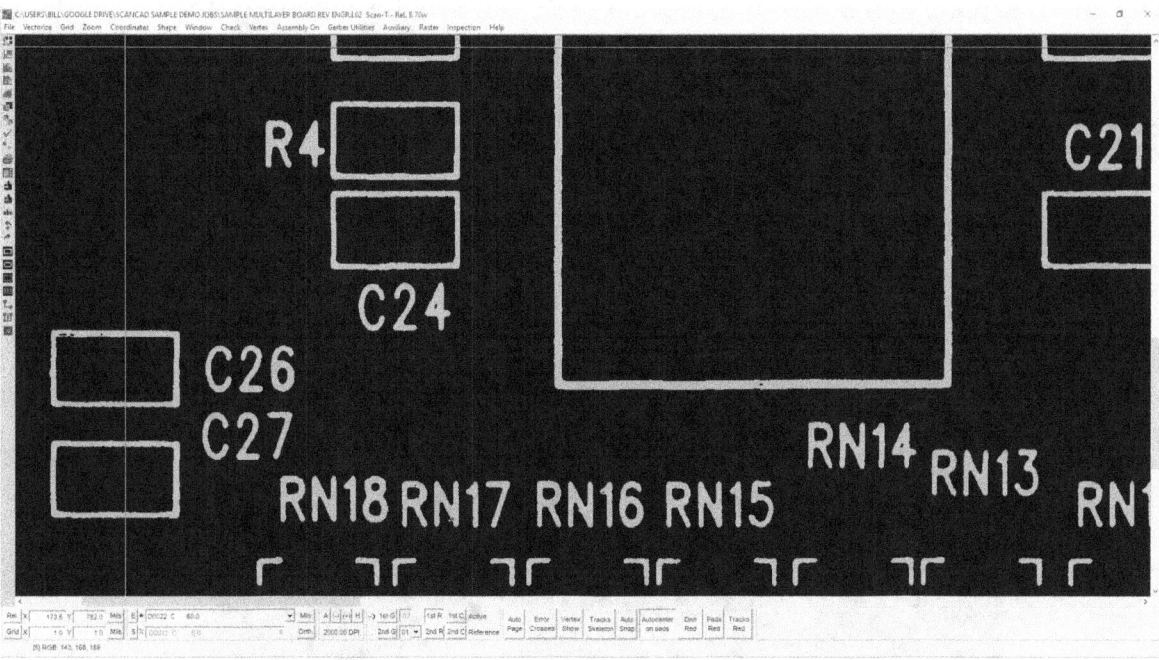

12. Gerber data after it was overlaid on monochrome legend raster image. Gerber data for this legend layer is now ready for output from ScanFAB.

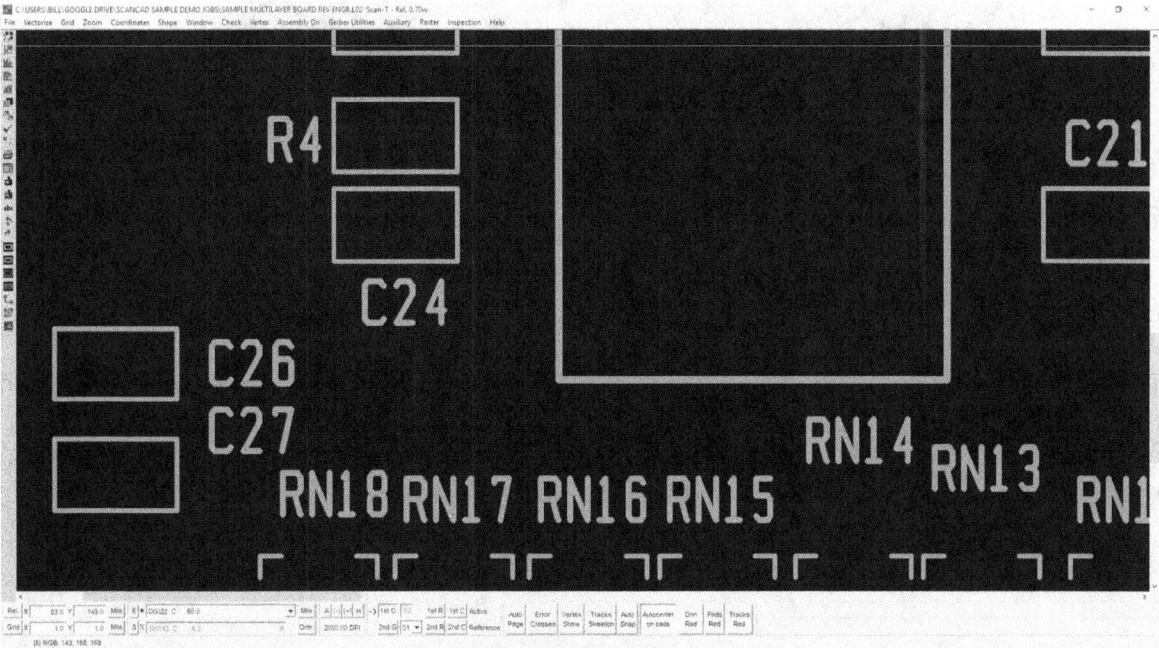

Tools & Techniques

The same process is repeated for every layer of the PCB: top, bottom, all inner layers, drill holes (both plated and non-plated) and board profile (route file). The resulting data will be form, fit and function Gerber, drill, route and BOM data for all layers of the PCB.

The resulting data from ScanFAB and ScanPLACE can then be imported into ConvertPLUS ARE where it is processed and component intelligence is added, such as footprints with pin numbering information, device designators, etc. ConvertPLUS ARE can then output data into a variety of intelligent netlist formats. These netlist formats can then be used to generate schematics, if necessary.

Below is a screenshot of ConvertPLUS ARE in action:[56]

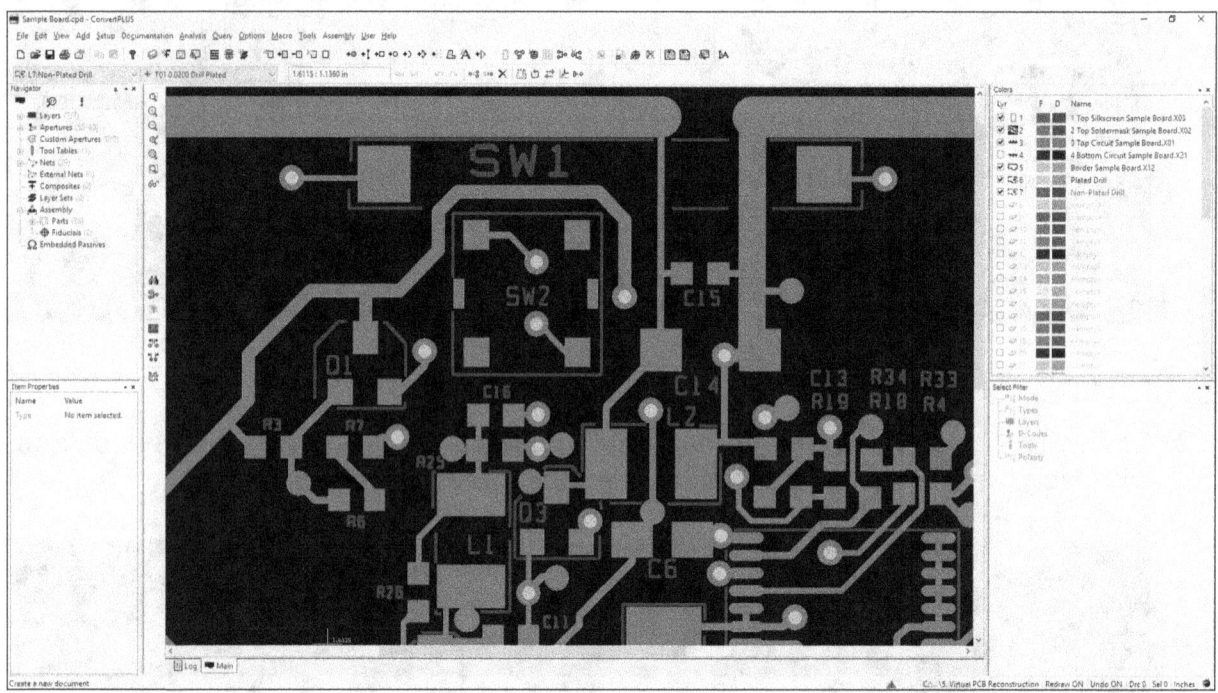

Endnote:

This addendum is put forth on request of the book's author, who felt that readers will benefit from real-life screenshots of ScanCAD in action to better appreciate the processes involved and the powerful PCB-RE features of this product. As the book is non-color, some of the details may not be apparent. Contact us and we'll be happy to arrange a demo.

---

[56] The image shows the Gerber data of a top circuit and solder mask layer of a PCB as well as the silkscreen and drill layer. This is the data before processing and adding component information.

Vendor Information

Equipment & Software

>Men have become the tools of their tools.
>
>Henry David Thoreau[57]

As incredible as it seems, engineers have come to rely more and more on the machines they built, not because they want to but because they need to. Technologies have long surpassed human abilities to handle the level of sophistication and complexity in product design and development without machine intervention. It goes without saying then, that ancillary tasks such as maintenance and repair of these systems will require the assistance of specialized equipment as well.

To some extent, PCB-RE by the manual approach is still possible but restricted only to certain level of PCBs that are manageable to the eyes and hands, as well as qualified engineers with the knowledge, skillsets and experiences to perform it. Even then, some form of visual and essential tools is still needed to circumvent inherent human limitations and weaknesses.[58] Beyond that, it's time to call in the big guns.

ScanCAD International, Inc.
P.O. Box 598
Morrison, CO 80465   USA
Phone: +1 303.697.8888
Fax   : +1 303.697.8580
E-mail: info@scancad.com
Web   : www.scancad.com

ScanCAD International, Inc. is an American corporation that provides a diverse variety of products and services to over 1000 companies in 47 countries. It has Headquarters located in Colorado, USA. Other offices include sales and support locations in Michigan & Florida, USA and two R&D Development locations in Europe.

---

[57] It's remarkable how this transcendentalist of the 19th century could have foreseen the day when man and machine become inseparable and interdependent on each other for mutual survival and existence.

[58] Engineers are mortals after all and they do succumb to visual impairment such as hyperopia and cataract, so proper lightings and magnifiers become indispensable as they grow older (and hopefully wiser).

Vendor Information

SCANCAD International, Inc.
Simplifying Complex Technology

As the name implies, ScanCAD offers a comprehensive suite of advanced scanning and software integrated systems. Behind its streamlined design and software architecture are a collection of modules that provide core functionalities, seamless integration and powerful management of different types of hardware, in-house or third-party supplied, that combines a dynamic blend of process management and control, automated optical inspection and measurement, electronic work instruction and manufacturing software (MES) that allows for the dissemination and collection of manufacturing intelligence (MI) throughout organizations.

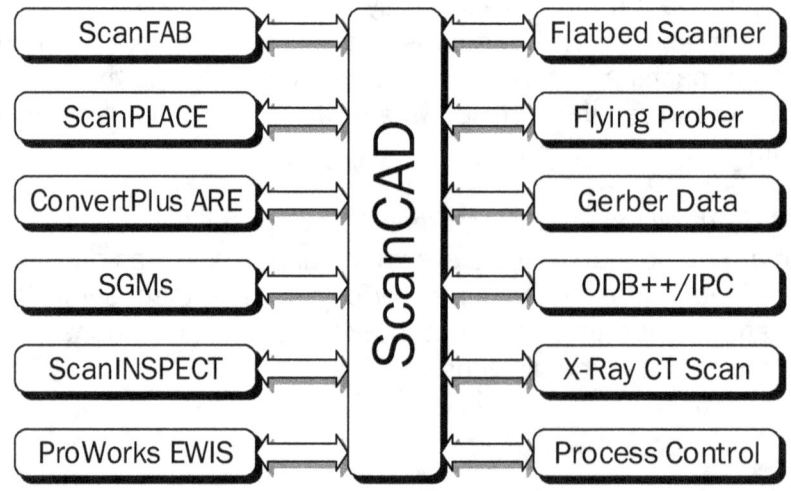

While ScanCAD offers an array of software products that are used in many engineering fields such as semiconductor, aerospace, medical, energy, automotive, construction, textile, military industries, etc., one of its remarkable strengths is in the PCB reverse engineering domain. In fact, the very first re-engineering systems were installed in 1987, even before the company was formally founded in 1990. Since then, ScanCAD products have been extensively used in over 1000 companies with a presence in 47 countries worldwide, and recognized as the number one ranked supplier of Electronic Work Instruction Software (EWIS) since 2009.

And as we shall see, ScanCAD is the only system in the world that can take an actual PCB and fully reverse engineer it back to complete manufacturing CAD data, with the ability to capture precise form, fit and function for all layers of different PCB types, with automatic layer-to-layer alignment and up to 99 layers per job. Comprehensive PCB deconstruction solutions are also available to choose from, whether destructive or non-destructive, as per the re-engineering requirements.

ScanFAB

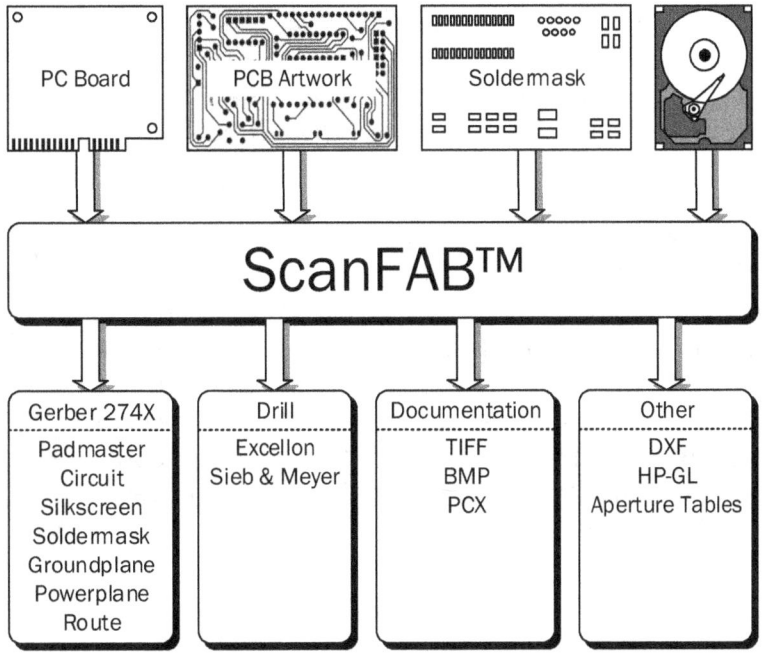

Linking existing designs to CAD and production.

ScanFAB is a fully integrated, stand-alone, scanner-based re-engineering system that permits the creation of CAD data (DXF/Gerber/Drill/CNC) from existing multilayer PCBs, parts, photo-tools, stencils, drawings, microfiche, PDF files, X-Ray images, etc. It also contains a full Gerber editor that can be used to import, modify and export Gerber & Drill data. As a Windows-based software that is linked to a high-resolution, calibrated flatbed scanner, this combination allows for accurate reverse engineering and precise reproduction of data to exact FORM, FIT and FUNCTION for today's high-density PCB board designs, complex parts and tooling.

The full FAB product has several powerful automatic vectorization functions that significantly reduce the time and energy required to re-engineer PCBs. The resulting Gerber formatted data files meet the high-quality standards that are required in the industry and is the reason why ScanFAB is the most successful and widely used PCB re-engineering system in the world today. And with the ability to import images from other imaging sources such as CT-Scan and X-ray machines, high-end optical imaging systems, Scanning Electron Microscopes (SEM) and more, even uncalibrated sources can even be overlaid on top of images acquired with the calibrated ScanCAD scanner from either side of the PCB, then stretched or scaled so that the additional imported images essentially become calibrated.

What follows is a discussion of the optional modules that can expand on the capability of the ScanFAB system.[59]

---

[59] Only functional descriptions are provided. For detailed specifications, refer to vendor's product datasheets.

Vendor Information

Precision Material Removal System

The Precision Material Removal System provides a very effective solution to delaminate PCBs by removing material in a very precise manner, one layer at a time, with minimal risk of damage to inner layers, avoiding lost data or incorrect data generation. The system is fast, easy to use, safe and provides very high-quality results.

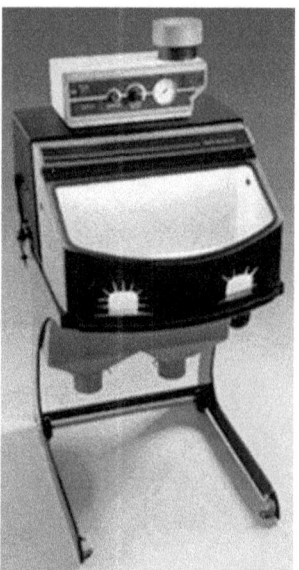

Standard system includes a turn-key package of custom processes, supplies and equipment, and comes with nozzles and abrasive material that is fine tuned for the PCB delamination application. The product is clean, quiet and safe, and fitted with an advanced HEPA filter built into the unit. Setup is easy with the necessary knowledge and training material (videos, workflows, etc.) provided to get the user up and running with the delamination process quickly. An optional ESD control version is also available for conformal coating removal of PCB's with static sensitive parts.

FPT Software Module

When a PCB needs to be re-engineered but cannot be sacrificed, the FPT Software Module integrates the ScanFAB software with the flying probe tester (FPT) to extract the netlist without destroying the PCB.[60]

For best results, when a PCB can be deconstructed, the FPT software module can work together with ScanFAB to permit 100% independent validation of a netlist, even if there is only one remaining PCB in existence. First, the FPT software module directs the flying prober machine to generate a netlist via probing and learning. ScanFAB then generates a second netlist independently during the destructive delayering process. The two resulting netlists can then be compared and reconciled to create an accurate, verified and validated netlist.[61]

The FPT Software Module is designed to interface with a wide variety of available flying prober machines. Not only does the software create the required test programs for the FPT, it also processes the error logs that are generated on the FPT to extract the netlist.

---

[60] The ScanFAB system can be used to program the test points for the FPT so it knows where to perform the electrical probing, which is the first step in the process of re-engineering with a FPT. Since CAD data is not available, this is provided by the ScanFAB system.

[61] Two independently created netlists from a single PCB dramatically reduces the risk associated with PCB-RE when only a single PCB is available to work on.

ScanPLACE

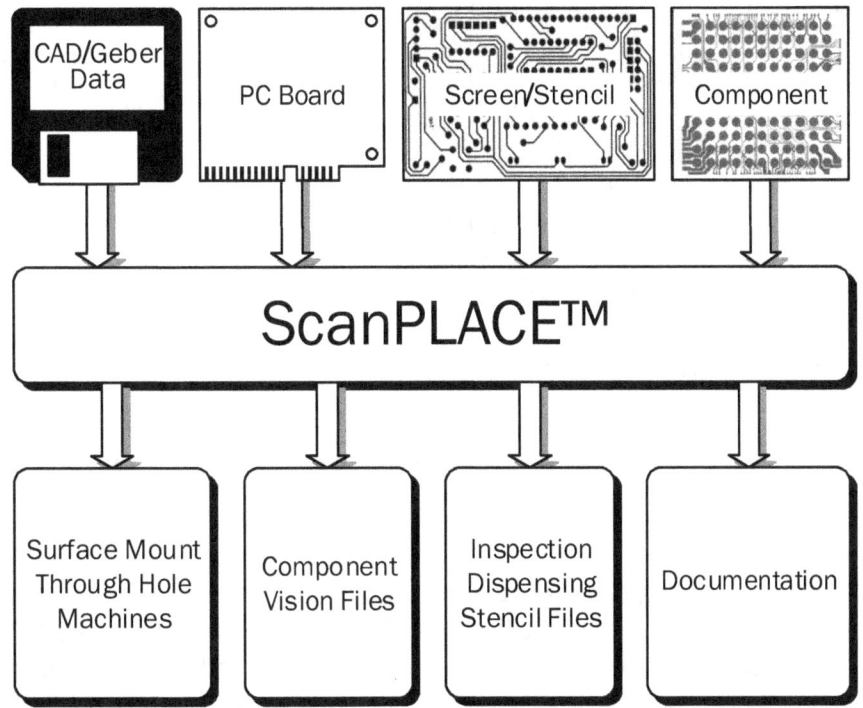

The ScanPLACE software module can be easily added to the ScanFAB system to provide the additional capability of extracting component centroid, rotation, package id, part number and reference designator information from the ScanFAB data, which can then be used to generate BOM information as well as assembly programs and process documentation for automatic PCB assembly systems such as surface mount component placement, through-hole component insertion, test, inspection and adhesive dispensing machines. This software module supports over 70 different output file formats. In addition, the software can be used to generate component vision file information for automatic placement equipment.

ConvertPLUS ARE

The ConvertPLUS ARE (automatic reverse engineering) software module takes the ScanFAB and ScanPLACE data and adds intelligence to the data such as component footprint and pin number information as well as extracting netlist information. Several intelligent netlist output file formats are supported such as ODB++, IPC-D-356A, IPC-2581 and more. These industry standard outputs can be imported into a variety of CAD packages for schematic generation or re-design.

Vendor Information

Schematic Generation Module

ScanCAD offers three options for schematic generation:

First, the Schematic Generation Lite Module, second, the full Schematic Generation Module with an output for a specific CAD software package and/or an intelligent PDF output and third, the ability to use one of the industry standard outputs such as ODB++ or an IPC format file from ConvertPLUS ARE that can be imported directly into a third-party CAD software package which can then be used for schematic generation.

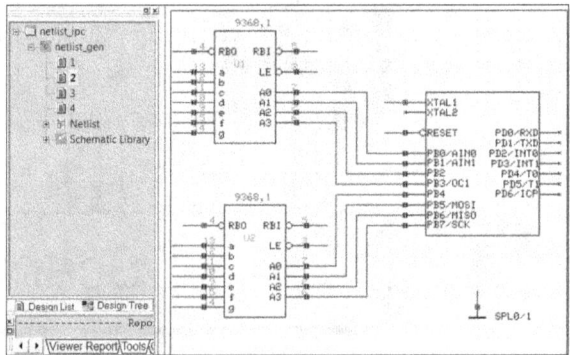

ScanINSPECT

It is also possible to add optional software to ScanFAB for first article measurement & inspection of artwork or even actual PCBs. The artwork or PCB can be 100% automatically inspected against Gerber data or PCBs can even be compared against each other (golden board inspection). This can be an important quality check step as part of the PCB re-engineering process or even a quality check of the re-manufactured PCBs that are manufactured based upon the Gerber data generated with the ScanFAB system. Compare a new PCB to the old PCB that was re-engineered to ensure they match exactly. ScanINSPECT uses the CAD data to inspect a PCB—it doesn't depend on an operator to 'know' what is supposed to be and not be present. It compares the PCB to CAD data to qualify the PCB as good or bad.

ProWorks Electronic Work Instruction Software

ProWorks is an interactive process control, or workflow software program that aids users with any function that requires work instructions, including performing a PCB re-engineering process. It provides the user with full documentation for each step needed to be performed including text, photos, videos, CAD files, supporting files and more. It has intelligent trouble-shooting and corrective action capabilities and recognizes when employees start to struggle on steps that are not getting passed immediately. At this point ProWorks will offer suggestions to the individual to help them solve the problem by themselves, without the need to bring in supervisors or tech support personnel. Employees are empowered and able to work more independently.

Appendix

References

> There are no answers, only cross-references.
>
> Norbert Wiener

The following references are related to PCB design and specifications:

INDUSTRY STANDARDS

Board Level

IPC-2221	Generic Standard for Printed Board Design
IPC-4101	Specification for Base Materials
IPC-4103	Materials for High Speed/High Frequency boards
IPC-7351	Surface Mount Design and Land Pattern
IPC-SM-840	Solder Mask Standard
IPC-TM-650	Test Methods Manual

Assembly Level

IPC-A-610	Acceptability of Printed Board Assemblies
J-STD-001	Requirements for Soldered Electrical and Electronic Assemblies

Documentation

MIL-STD-100	Engineering Drawing Practices
ASME Y14.100	Replaces MIL-STD-100 for non-Military
ASME Y14.5	Dimensioning and Tolerancing
IPC-D-325	Documentation Requirements for Printed Boards, Assemblies and Support Drawings

Appendix

PCBs and assemblies are grouped according to the following classes and types:

PERFORMANCE CLASSES

Class 1 – General Electronics Products

- Functionality is key. Aesthetics is secondary
- Computer systems and peripherals

Class 2 – Dedicated Service Systems

- Extended and uninterrupted service desirable but not required
- Communications, business machines, instruments

Class 3 – High Reliability Systems

- Continued or performance on demand is important. Downtime is not acceptable
- Life support systems, critical weapons and defense systems

BOARD TYPES

Type 1	Single-Sided Printed Board
Type 2	Double-Sided Printed Board
Type 3	Multilayer without blind and/or buried vias
Type 4	Multilayer with blind and/or buried vias
Type 5	Multilayer Metal Core without blind and/or buried vias
Type 6	Multilayer Metal Core with blind and/or buried vias

ASSEMBLY CLASSES

Class A	Through-hole mounted components only
Class B	Surface Mount (SMT) components only
Class C	Simplistic through-hole and SMT intermixed assembly
Class X	Complex intermixed assembly (TH,* SMT, fine pitch and BGA)
Class Y	Complex intermixed assembly (TH, SMT, ultra-fine pitch and chip scale)
Class Z	Complex intermixed assembly (TH, ultra-fine pitch, COB, flip-chip and TAB)

* TH – Through-hole

References

The Association Connecting Electronics Industries (IPC[62]) defines the following as guidelines for standardizing assembly and production requirements of electronic products:

General Documents

IPC-T-50	Terms and Definitions
IPC-2615	Printed Board Dimensions and Tolerances
IPC-D-325	Documentation Requirements for Printed Boards
IPC-A-31	Flexible Raw Material Test Pattern
IPC-ET-652	Guidelines and Requirements for Electrical Testing of Unpopulated Printed Boards

Design Specifications

IPC-2612	Sectional Requirements for Electronic Diagramming Documentation (Schematic and Logic Descriptions)
IPC-2221	Generic Standard on Printed Board Design
IPC-2223	Sectional Design Standard for Flexible Printed Boards
IPC-7351B	Generic Requirements for Surface Mount Design and Land Pattern Standards

Material Specifications

IPC-FC-234	Pressure Sensitive Adhesives Assembly Guidelines for Single-Sided and Double-Sided Flexible Printed Circuits
IPC-4562	Metal Foil for Printed Wiring Applications
IPC-4101	Laminate Prepreg Materials Standard for Printed Boards
IPC-4202	Flexible Base Dielectrics for Use in Flexible Printed Circuitry
IPC-4203	Adhesive Coated Dielectric Films for Use as Cover Sheets for Flexible Printed Circuitry and Flexible Adhesive Bonding Films
IPC-4204	Flexible Metal-Clad Dielectrics for Use in Fabrication of Flexible Printed Circuitry

---

[62] It was founded in 1957 as the Institute for Printed Circuits. Its name was later changed to the Institute for Interconnecting and Packaging Electronic Circuits to highlight the expansion from bare boards to packaging and electronic assemblies. (Source: Wikipedia)

Appendix

(continue...)

Performance and Inspection Documents

IPC-A-600	Acceptability of Printed Boards
IPC-A-610	Acceptability of Electronic Assemblies
IPC-6011	Generic Performance Specification for Printed Boards
IPC-6012	Qualification and Performance Specification for Rigid Printed Boards
IPC-6013	Specification for Printed Wiring, Flexible and Rigid-Flex
IPC- 6202	IPC/JPCA Performance Guide Manual for Single- and Double-Sided Flexible Printed Wiring Boards
PAS-62123	Performance Guide Manual for Single & Double Sided Flexible Printed Wiring Boards
IPC-TF-870	Qualification and Performance of Polymer Thick Film Printed Boards

Flex Assembly and Materials Standards

IPC-FA-251	Assembly Guidelines for Single and Double Sided Flexible Printed Circuits
IPC-3406	Guidelines for Electrically Conductive Surface Mount Adhesives
IPC-3408	General Requirements for Anisotropically Conductive Adhesives Films

References

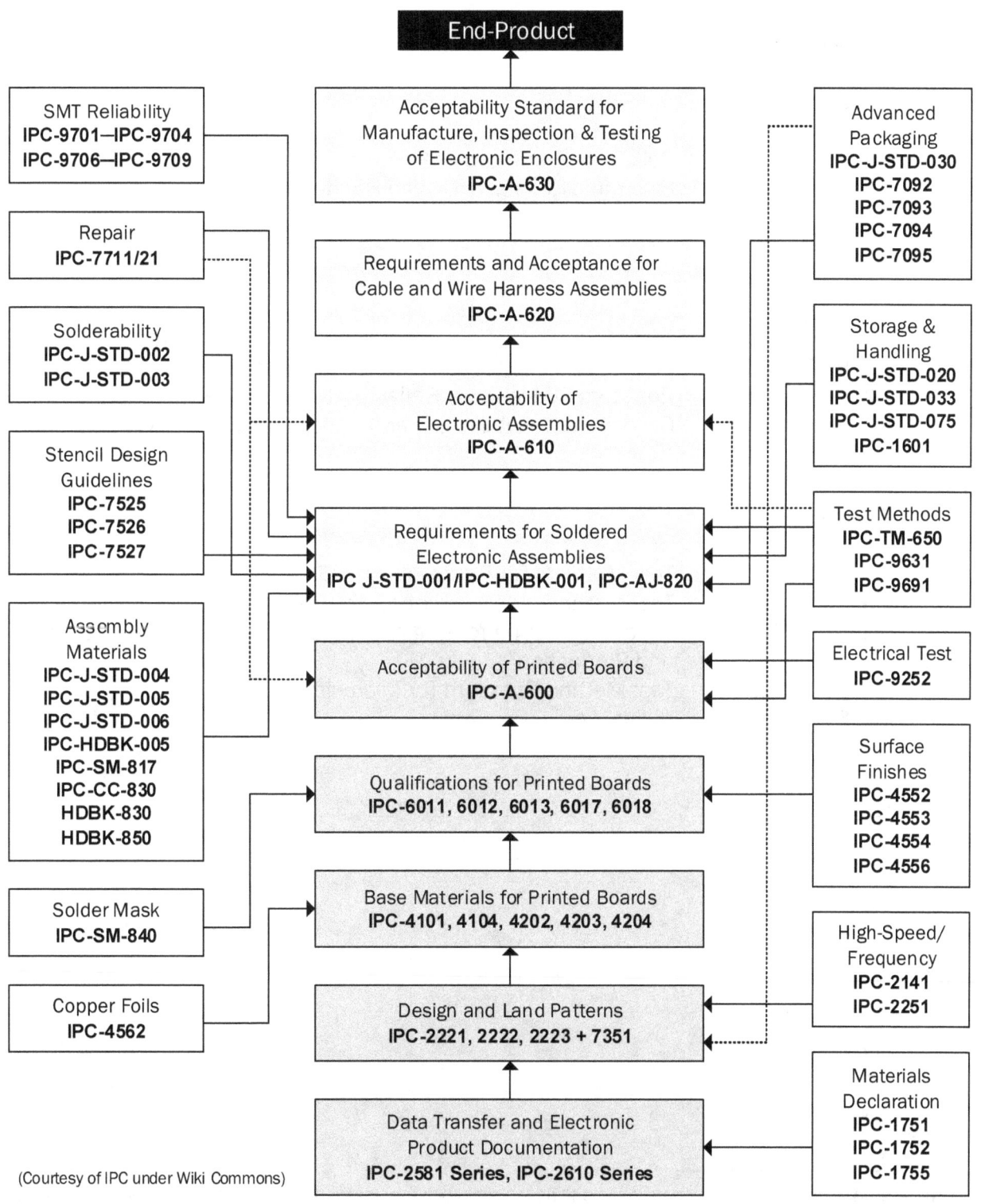

(Courtesy of IPC under Wiki Commons)

Appendix

US military standards[63] related to electronics are listed below:

MILITARY STANDARDS

Handbooks

MIL-HDBK-271F	Reliability Prediction of Electronic Equipment
MIL-HDBK-338B	Electronic Reliability Design Handbook

Performance

MIL-PRF-19500	Test Methods for Semiconductor Devices, Discretes
MIL-PRF-38535	Integrated Circuits (Microcircuits) Manufacturing
MIL-PRF-55342	Resistors Space-Level Reliability General Specifications
MIL-PRF-55681	Capacitors Reliability General Specifications

Standards

MIL-STD-202	Test Methods for Electronic and Electrical Parts
MIL-STD-750	Test Methods for Semiconductor Devices
MIL-STD-781D	Reliability Testing for Engineering Development, Qualification and Production
MIL-STD-883B	Test Method Standard for Microcircuits
MIL-STD-1553	Multiplexed Data Bus

---

[63] These standards underwent various revisions as technologies and methodologies improved overtime, so some are intentionally left out to keep the references generic.

References

The following references are ISO[64] standards related to electronics:

ISO STANDARDS

ICS[65]

31.020	Electronic Components in General
31.040	Resistors
31.060	Capacitors
31.080	Semiconductor Devices
31.120	Electronic Display Devices
31.140	Piezoelectric Devices
31.160	Electric Filters
31.180	Printed Circuits and Boards
31.190	Electronic Component Assemblies
31.200	Integrated Circuit, Microelectronics
31.220	Electromechanical Components (Electronic & Telecommunications)
31.240	Mechanical Structures for Electronic Equipment
31.260	Optoelectronics, Laser Equipment

ISO[66]

9001	Quality Management Systems (QMS)
9002	QMS for Manufacture and Delivery of Products
9003	QMS for Inspection and Testing of Products

---

[64] Commonly but incorrectly mistaken as International Standards Organization. The proper term is International Organization for Standardization (www.iso.org).

[65] The International Classification for Standards (ICS) is a convention managed by the International Organization for Standardization (ISO) and used in catalogues of international, regional, and national standards and other normative documents. The latest edition of the ICS can be downloaded free of charge from the ISO web site.

[66] When they were first introduced, a company would decide which of the standards they should be certified to: ISO 9001 for design and production, ISO 9002 for production or ISO 9003 for inspection and testing, depending on which type of industry the company was in or what the company did. From 2000 onwards, ISO 9002 and 9003 are no longer used.

Appendix

ABOUT THE AUTHOR

NG KENG TIONG is an engineer turned writer with a passion to share his knowledge and experience of over 30 years in electronics in the field of PCB-RE, testing and repair. He formerly worked as a Principal Engineer at Singapore Technologies (ST) Electronics Limited, a subsidiary of ST Engineering. Upon graduation from the Singapore Polytechnics, he signed up with the Republic of Singapore Air Force (RSAF) as an aircraft technician and worked in the E-2C Hawkeye repair bay, servicing the aircraft's avionics using automated test systems (CAT-IIID and RADCOM) and in-house test equipment.

Upon invitation, he left the RSAF after his first contract and joined the home-grown defense industry, writing test programs and doing PCB diagnostics on Schlumberger S700 series testers. He had worked on other test platforms such as the Teradyne Spectrum 8800 series, the Westest DATS/2000 test station, and some special-to-type-equipment (STTE) of similar nature. He also has experience in logic simulation using the HHB Systems CADAT software and CATS-10000 hardware modeler, as well as Teradyne's LASAR simulator.

Through the course of his work, he encountered many printed circuit boards and electronic modules without schematic diagrams or documentation. That started him on the journey of doing PCB reverse engineering, in part or total, to perform the necessary troubleshooting for repair. Over time, he has refined the skill into an art and re-produced it into a book (see overleaf).

PCB reverse-engineering (PCB-RE) is a skill that requires more than just a passing acquaintance with electronics. To the uninitiated, it is a difficult if not impossible undertaking reserved only for the determined and qualified. The author, however, believes that armed with a right mindset and equipped with the right knowledge will enable even the average electronics engineer to do it.

If you are interested to learn how to do PCB-RE using the manual approach, this book will teach you the steps and guide you along, using Microsoft Visio as the tool of choice to document the process.

Visit **visio-for-engineers.blogspot.sg** for more information and special discount offer!

Complimentary copies are freely available to customers of ScanCAD International, Inc. Contact ScanCAD today to secure your copy!

Reviews for the Book

Having spent a few weeks working through the book, I can pretty much confirm that this is the best resource out there for the subject. The book not only discusses the concepts of reverse engineering, but delves extensively into using Visio as a tool to neatly capture the process as well, something I wouldn't have considered had it not been for this book delving into such details. Well done to the author, and this book will serve as a constant companion during my future RE activities.

Gert Byleveldt
Specialist in Automotive Electronic Repairs

I purchased this book after having discussed it with the author on the EEVBlog web forum. I have been reverse engineering complex printed circuit boards for many years, and enjoy the challenge. The author of this book has similar experience and it shows in his excellent coverage of the topic. He has correctly identified the key aspects of such a task and the need for the methodical and disciplined approach to achieve success.

This book would suit anyone who has an interest in the reverse engineering of PCB's, and even those with previous experience will likely learn from this book. I certainly did! There is no requirement for complex or expensive tools beyond the basics normally found on an electronics workbench, but MS Visio is recommended for the documentation. Visio is not essential to success however, and should not put buyers off this book.

I know of no other book that covers this topic so well for those starting out on PCB reverse engineering. The author writes in an easy to read fashion and offers purchasers of his book free downloads of additional useful material via his web site. The quality of materials used in the production of this book are excellent and it should have a long life in the lab. All who have seen my copy of the book have been very impressed with it. Highly recommended for those embarking on PCB reverse engineering, both beginners and the experienced.

Fraser Castle
Electronics engineer and hobbyist
Amazon.co.uk

Literature on reverse engineering (i.e. the analysis and possibly reconstruction of a finished system) is rare, and the few books devoted to this topic are usually so general that every aspect can be treated only briefly. The book *The Art of PCB Reverse Engineering* by Ng Keng Tiong therefore is an exception: it is dedicated to the reverse engineering of electronic circuits and exclusively deals with this subject in detail.

The author deals professionally for about 15 years with the reconstruction and repair of electronic systems. In his book, he summarizes the experience gained in a structured manner and apply the methodology progressively using an example PCB——a small ISA bus SCSI host adapter. Each step of the reverse-engineering process, from identification of elements, analysis of electrical connections, reconstruction of the circuit diagram, is treated in a separate chapter; numerous illustrations and repeatedly interspersed anecdotes of the author from his own professional activity make the book enjoyable to read and easy to understand.

To document the information obtained the author uses the Microsoft Visio program. Engineers who use this program will therefore benefit from the detailed step-by-step guide. However, a chapter dedicated to both commercial and open source EDA programs also addresses readers who use other software tools or are still looking for such tools.

Overall, it's a very interesting and unusual book for anyone working professionally or as a hobby in the analysis of electronic systems——whether to repair them despite the lack of circuit documents, but if only to understand how they work and as the subtitle of the book states——to rediscover the beauty of their original designs.

Dr. Stephan Pabst
ETAS GmbH
Engineering Services Automotive (ETAS/PGA-EAS2)
Borsigstraße 14
70469 Stuttgart
Germany
www.etas.com

Just received my copy of your book from Amazon. Amazing! This is going to be one of the most useful books I have ever bought. Visio has always been one of my favorite tools, and I would consider myself quite an expert, but you have shown me some wonderful new tricks!

Ken Howard
Configuration Manager
Wide Area and Space Surveillance Systems Program Office
Department of Defense | Capability Acquisition and Sustainment Group

I just want to thank you for your amazing book. I really enjoyed reading it. I am in reverse engineering field for some time but have never used Visio. It was an amazing idea. I know your hard work will not pay back financially but at least you will have support from people like me and I will buy anything you publish in the future. I've left my review on Amazon.ca and I will recommend your book to all my colleagues I am working with in Canada.

Amir Pasalic, B. Sc.EE
ENA Electronics Inc.

Just to let you know that I bought your book via eBay, just the standard edition as I'm color blind anyway. I finished reading it in a few sittings over two days. I'm 65 years old and now retired. As a youth I loved repairing things, then I became an electronic design engineer and now that I'm retired I am loving repairing things again. It's just like reading a good detective novel!

I can see that you really put your heart into this book and it is really nicely presented. I have done some haphazard reverse engineering previously for repairs but I have found your excellent book to be a good lesson in applying method and discipline to the process.

Regards
Bob Dring
Sydney, Australia

Art without engineering is dreaming. Engineering without art is boring. §

                       Steven Roberts

---

§ The original quote is 'Engineering without art is calculating', but 'boring' seems to fit better, as engineers often give people the impression that they are a boring lot with nothing else to do but solve problems.

www.ingramcontent.com/pod-product-compliance
Lightning Source LLC
Chambersburg PA
CBHW082341220526
45470CB00008B/2589